T0183301

SpringerBriefs in History of Science and Technology

Series Editors

Gerard Alberts, University of Amsterdam, Amsterdam, The Netherlands

Theodore Arabatzis, University of Athens, Athens, Greece

Bretislav Friedrich, Fritz Haber Institut der Max Planck Gesellschaft, Berlin, Germany

Ulf Hashagen, Deutsches Museum, Munich, Germany

Dieter Hoffmann, Max-Planck-Institute for the History of Science, Berlin, Germany

Simon Mitton, University of Cambridge, Cambridge, UK

David Pantalony, University of Ottawa, Ottawa, ON, Canada

Matteo Valleriani, Max-Planck-Institute for the History of Science, Berlin, Germany

More information about this series at http://www.springer.com/series/10085

Alexei Kojevnikov

The Copenhagen Network

The Birth of Quantum Mechanics
from a Postdoctoral Perspective

 Springer

Alexei Kojevnikov
Department of History
University of British Columbia
Vancouver, BC, Canada

ISSN 2211-4564   ISSN 2211-4572   (electronic)
SpringerBriefs in History of Science and Technology
ISBN 978-3-030-59187-8   ISBN 978-3-030-59188-5   (eBook)
https://doi.org/10.1007/978-3-030-59188-5

This Springer imprint is published by the registered company Springer Nature Switzerland AG
The registered company address is: Gewerbestrasse 11, 6330 Cham, Switzerland

In memory of my teacher,
Igor Serafimovich Alekseev (1935–1988),
philosopher and Marxist interpreter of
Bohr's complementarity

Acknowledgements

This project was supported by the Max-Planck-Institut für Wissenschaftsgeschichte, Berlin, and by the SSHRC grant 410-2011-0936. I wish to thank all members of the quantum history group at the institute, in particular Jürgen Renn, Giuseppe Castagnetti, Hubert Goenner, Dieter Hoffmann, Edward Jurkowitz, Horst Kant, and Arne Schirrmacher, for many productive discussions. My special gratitude goes to Finn Aaserud, Christian Joas, Felicity Pors, and the staff of the Niels Bohr Archive in Copenhagen for their hospitality and invaluable advice. For further comments and suggestions, I am thankful to Richard Beyler, Cathryn Carson, Olivier Darrigol, Paul Forman, Olival Freire Jr., John L. Heilbron, Anja Skaar Jacobsen, Diana Kormos Barkan, Helge Kragh, Dorinda Outram, Ted Porter, Helmut Rechenberg, Eric Scerri, Skuli Sigurdsson, and participants at the colloquia at Deutsches Museum, Munich; Institute for History of Science and Technology, Moscow; Harvard University; University of California Los Angeles; and Universidade Federal da Bahia. I am grateful to Lindy Divarci and Stefanie Ickert for editorial assistance. An earlier version of Chap. 6 was published in (Kojevnikov 2011).

Contents

Abbreviations and Archives

AHQP	Archive for the History of Quantum Physics, APS and many other depositories
AP	*Annalen der Physik*
APS	American Philosophical Society Library, Philadelphia
BCW	Bohr *Collected Works* (Bohr 1972–2008)
BGC	Bohr General Correspondence, NBA
BKS	Bohr–Kramers–Slater theory
BSC	Bohr Scientific Correspondence, NBA
DM	Deutsches Museum Archive, Munich
ETH	ETH Library, Zurich
HDQT	*The Historical Development of Quantum Theory* (Mehra and Rechenberg 1982–2001)
HSP(B)S	*Historical Studies in the Physical (and Biological) Sciences*
IEB	International Education Board
IMN	*Intellectual Mastery of Nature* (Jungnickel and McCormmach 1986)
NBA	Niels Bohr Archive, Copenhagen
NW	*Die Naturwissenschaften*
PWB	Pauli *Wissenschaftlicher Briefwechsel* (Pauli 1979)
PZ	*Physikalische Zeitschrift*
RAC	Rockefeller Archive Center, Sleepy Hollow, NY
SWB	Sommerfeld *Wissenschaftlicher Briefwechsel* (Sommerfeld 2000–2004)
ZP	*Zeitschrift für Physik*

Unless otherwise stated, cited archival documents and letters are microfilm copies from AHQP; Bohr's correspondence from BSC.

Chapter 1
Introduction

Quantum mechanics will soon be one hundred years old and still has not been disproven. This fact would have certainly surprised many of its creators, the physicists who lived in an era of great social and scientific turmoil, when firmly established ideas were being overturned. Many of the very authors who gave us quantum mechanics believed that it, too, would soon be succeeded and superseded by another, even more radical and fundamental breakthrough, and tried to make this happen by calling into question the very foundations of their own achievement. The historiography of quantum theory has thoroughly analyzed the development of scientific ideas and concepts. In comparison with intellectual history, much less has been done with regard to the historical understanding of broader cultural meanings and social context of the quantum revolution in twentieth-century science, a growing enterprise to which this book, as well as other books in this series, aims to contribute.[1]

Using approaches from the cultural and social history of science, this study focuses primarily on the nascent quantum mechanical community of physicists—predominantly a young, multinational, and multicultural group—consisting mainly of postdoctoral and graduate students. During economically uncertain times, they typically survived on temporary appointments, fellowships, and "soft money," traveling and conducting research at various places. Niels Bohr's Institute for Theoretical Physics in Copenhagen, where many of them spent time as visitors, postdoctoral fellows, or

---

[1] This book is the second in the planned series of four volumes addressing the beginnings of quantum physics research at the major European centers. The first investigation (Schirrmacher 2019) focused on Göttingen; the current one deals with the Copenhagen network, and the two subsequent ones will study the centers in Berlin and Munich. These investigations emerged from an expansive study on the quantum revolution as a major transformation of physical knowledge undertaken by the Max Planck Institute for the History of Science and the Fritz Haber Institute (2006–2012).

It is impossible to do justice to the incredibly rich conceptual historiography of quantum physics. Only the most general surveys range from the early and still very valuable classics (Jammer 1966; Hund 1974) to the encyclopedic six-volume work by Mehra and Rechenberg (1982–2001) and to the most recent analysis in Duncan and Janssen (2019). More specific studies will be cited further in the text. The pioneering approaches in, respectively, the social and cultural history of quantum physics originate from Forman (1967, 1971).

© The Author(s), under exclusive license to Springer Nature Switzerland AG 2020
A. Kojevnikov, *The Copenhagen Network*, SpringerBriefs in History of Science
and Technology, https://doi.org/10.1007/978-3-030-59188-5_1

conference participants, served as the symbolic geographical center and the clearing house of ideas for this emergent disciplinary community. This informal Copenhagen network was primarily responsible for the invention of quantum mechanics in 1925–1927 and the almost immediate spread of the new theory worldwide.

1.1 *KnabenPhysik*

It is, of course, well known that many of the pioneers of quantum mechanics were quite young when they created their revolutionary theory. To illustrate this point, in the year-and-a-half after the first draft proposal by Heisenberg in July 1925, more than 200 contributions to the new quantum theory appeared, mostly articles, thanks to the exceptionally fast rate of journal publications in that period, but also a few books. Over 80 authors took part in that brainstorming effort: The majority of them were under 30 years of age and they authored almost 70% of all publications.[2] Some were still working on their dissertations, but more commonly, they were recent PhDs, having obtained their degrees after 1920, and would have been considered postdoctoral students by today's standards. My referring to this group as "postdoctoral," in a generalized sense, requires clarification that the term was much less commonly used at the time. It was still new and relatively unfamiliar, at least in Europe, a recent importation from the USA. The new quantum mechanics, as we shall see in this study, would become one of the unexpected intellectual outcomes of that novel institutional development.

If we become more selective and focus only on the most important creators of quantum mechanics, the postdoctoral category, understood in the above broad meaning, would include 5 out of 9, or 8 out of 12 major contributors, depending on the count. And, if we permit ourselves to be completely elitist and judge by the standards of the Nobel Committee, then two out of three theoretical physicists who received their prizes for, as was generally understood then, quantum mechanics in 1932 and 1933 were young and recent PhDs: Werner Heisenberg (born 1901, PhD 1923) and P.A.M. Dirac (born 1902, PhD 1926). The third, Erwin Schrödinger (born 1887, PhD 1910), was much more mature physically and professionally in comparison with the other two, if not with an average Nobel laureate, and participated in the creation of quantum mechanics already as a full professor approaching forty. The photograph of all three men taken at a railway station in Stockholm in December 1933 during the joint Nobel ceremony further emphasizes this point if we consider its oft unnoticed gendered aspect. The Nobel ritual allotted a special role to the laureate's partner, so each was expected to come with a female companion. Yet only Schrödinger was able to bring his wife along, as both Heisenberg and Dirac were not only unmarried, but did not even have girlfriends, as far as their biographers can

---

[2]These numbers are based on more detailed statistics and the bibliography of early quantum mechanics in Kozhevnikov and Novik (1989).

Fig. 1.1 1933 recipients of the Nobel Prize in physics *Source* Tekniska Museet, Stockholm

tell us. The photograph thus shows both of them arriving for the Nobel award in the company of their mothers (Fig. 1.1).[3]

Not just in this one episode, but generally, the youthfulness and immaturity of the many contributors to quantum mechanics were the topic of frequent jokes at the time, and quantum mechanics itself was sometimes, jokingly and colloquially, referred to as *Knabenphysik*, or "teenagers' physics." Yet as the saying goes, in every joke there is only a portion of a joke, and in this book, I will explore the more serious aspects and consequences, both historical and philosophical, of *Knabenphysik*. The first important corollary for the intellectual history that follows is that we should not automatically assume that the main actors in the story were independent, both socially and in terms of the scientific work they did. What did it mean for knowledge production if the typical author was a young scientist with postdoctoral status or similar? First and foremost, the postdoctoral position was then, even more so than it is now, a subordinate and precarious one, lacking security and long-term predictability. Its holders could be up-and-coming, but still unsure about their ultimate career prospects. Often very bright and having important ideas of their own,

---

[3]On this point, Schrödinger's biographers could have commented that his sexual experiences abundantly compensated for the lack of Heisenberg's and Dirac's and that he certainly would have been willing to bring a few additional female companions. On Schrödinger's personal life and science, see Moore (1989).

they were still not fully emancipated financially, institutionally, and also often in their research, needing to adapt to the power and strategies of the senior professors, on whose patronage they relied.

The above is true even in stable and prosperous times, but the short period in the 1920s could be considered relatively stable only in comparison with the extraordinary disruptions of the preceding and the following decades. The interwar period continued to suffer from the lingering aftereffects of the Great War, from social and political conflicts that would ultimately lead to the next global conflagration, and from repeated economic crises and financial meltdowns, including the 1923 hyperinflation and the 1929 stock collapse. The academic profession was hit by the depression particularly hard, with drastically decreased job prospects that forced many talented and aspiring students to look for more realistic careers outside of universities. In part to alleviate their plight, postdoctoral fellowships were introduced to Europe in the 1920s. Even the most successful young academics, such as Heisenberg and Wolfgang Pauli, suffered from uncertainty during this period. It requires some effort, psychologically, to think about the latter two as vulnerable and insecure students, without permanent positions and funding, given their meteoric rise to fame and subsequent status as great scientists. But once we make this effort of historical empathy, we are better able to understand and reinterpret some of their important moves, choices, and ideas for the radical reconceptualization of quantum theory during the 1920s. One of the main conclusions of this study is that a number of important and familiar episodes in the history of quantum mechanics appear in a rather different light if we abandon the idealization of an independent great mind—typical in intellectual history—and instead assume the perspective and the agency of a precarious postdoctoral student.

It had not happened before in the history of science that such a large group of young scholars coalesced so quickly around an emerging new discipline or approach and provided a critical mass for its development. Unlike relativity theories, either special or general, or the "old quantum theory," the new quantum mechanics of 1925–27 immediately became a large collective enterprise. For the scale of scientific research in the 1920s, which was much smaller than today's, 200 publications in 18 months were an unprecedented phenomenon, and the initial group of roughly 80 authors from 14 countries must have certainly doubled or tripled during the following year. One factor behind such unstoppable momentum was the already mentioned speed of publications: It was not uncommon at the time that a paper submitted to a journal would be published in approximately two to three months and about one month later would be already cited in a paper submitted for publication by another author. Yet formal publications alone do not explain such explosive growth, which could not have happened without frequent personal contacts, travels, hundreds of letters, and informal exchanges of proof sheets of as yet unpublished articles. Unlike a typical nineteenth-century pattern of discipline formation, in the case of quantum mechanics, no single major center or institution of graduate training could accommodate such a large community of researchers. The latter's geographical spread and unprecedented mobility enabled informal exchanges between remote locations. Its members often pushed the work in diverging, sometimes contradictory directions, so that no individual leader could stay effectively in charge or claim ultimate credit for the enterprise. I will explore further philosophical consequences of such a dispersed mode of new discipline creation in Chap. 6 of this book.

The itinerant nature of postdoctoral positions contributed to their holders' experiences of marginality combined with opportunity. Their employment was not only short term but required leaving familiar local surroundings and home institutions, perpetuating greater flexibility in learning new and changing research agendas. Most postdoctoral fellowships at the time formally required going to a foreign country, sometimes to more than one, which made the resulting community of quantum postdocs international and to some degree extraterritorial. As we will see, the transitory lifestyle allowed some of these students a modicum of freedom to develop independent conceptual agendas that sometimes contradicted those of their professors. Decades later, they would nostalgically recall the internationalist spirit of the 1920s, but we should not take these fond memories about youthful days too literally. Science in the 1920s still remembered the nationalistic convulsions and ideological wounds of World War I, and it was still virtually impossible for a German or an Austrian scientist to travel to France, Belgium, or Britain, and vice versa, until the latter half of that decade, and even then, only rarely. A relatively short opening toward limited internationalism would start closing again in the 1930s, as Europe braced itself for another war. By the standards of the time, the scientific internationalism of the quantum mechanical community was an exception rather than the norm, and one of the goals of this study is to explore what made this social and political phenomenon possible.

Yet this international community had a geographical center, at least in a symbolic sense, but the center was located in one of the smallest European countries that, in normal conditions, would have been unable to compete with more established scientific powers in the amount of funds provided for research, in career opportunities for young scientists, or in training students. Thus, another major problem for this study is to explain what combination of resources allowed Copenhagen, for nearly a decade, to function as the international "Mecca of theoretical physics."

1.2 "Mecca of Theoretical Physics"

Insiders' and outsiders' perspectives often present revealing contrasts. Take, for example, the Nobel Prizes awarded for glorious discoveries in physics during the revolutionary early half of the twentieth century. From the point of view of the outside world, toward the end of each year, the public waited anxiously to hear news of which physicist would be admitted as the next member to the club of immortals. From places as remote as Berlin, Paris, Cambridge, Moscow, and Pasadena, they expected the Stockholm Areopagus to exercise its highest authority and issue a verdict on the discipline's most important achievement. A historian who enters the archives of the Nobel Committee and reads its almost century-old records, encounters a very different Scandinavian perspective. The five local physicists comprising the committee were not necessarily connected to the faraway places where crucial scientific advances were taking place, and consequently, not always sure about their ability to confidently evaluate sophisticated developments at the forefront of physics, especially in the fields outside their areas of direct expertise. Their major concern was to avoid making a damaging mistake that could undermine the reputation of the prize, and the committee often preferred a safe and solid accomplishment to one that was possibly much more spectacular and exciting, but risky.[4]

Although the insiders' worries were somewhat exaggerated, it is nevertheless true that the international prestige of the Nobel Prize could not be taken for granted during the early decades of its existence. To name just one problem, the Swedish Academy's claim to serve as the impartial arbiter and the highest international authority on principal advances in world science had to be accepted by larger, more productive, and much more arrogant scientific powers. Correspondingly, the academy and the Nobel Committee saw as their primary, but neither easy nor guaranteed objective, to win such recognition for their small country and establish the international reputation of the Nobel award as the unquestionable hallmark of most important scientific achievements. Eventually, they succeeded in this task remarkably well, as we all know. The prestige of the prize became so firmly entrenched in the decades following World War II that it literally began to be taken for granted; today, its status appears to us so natural that is rarely seen as a historical problem. Only some Scandinavian historians, better familiar with insiders' perspectives, occasionally wonder how it became possible for Sweden to establish itself in the scientific world as a "center on the periphery" (Lindqvist 1993).

The ascendancy of Copenhagen as the world center of theoretical physics in the 1920s is also rarely problematized. In March 1921, the recently appointed professor Niels Henrik David Bohr officially inaugurated the building of the newly established Institute for Theoretical Physics at the University of Copenhagen. The modest three-story facility included a small experimental laboratory and had, in addition to the director, three staff positions for an assistant, a mechanic, and a secretary. Five years later, the institute had become the destination of scientific pilgrimages by theoretical physicists from all over the world and the focus of an international network that created a fundamental scientific breakthrough, quantum mechanics. Over 60 long-term researchers worked there during the 1920s, along with many more short-term visitors, who together made crucial contributions to the revolutionary new science and spread it worldwide. Once again, an insider's perspective and awareness of the smallness of both the Niels Bohr Institute and the Danish resources for science are required to make one wonder what made this remarkable success story possible in Copenhagen—a city that had not previously been known as a significant location of physics research.

A historical answer to the above question requires much more than simply pointing out the greatness of Bohr's ideas and the attractiveness of his personality. These usual

---

[4]This is largely the reason why, for example, Albert Einstein received his Nobel Prize not for the theory of relativity, either special or general, but for the empirically confirmed formula of the photoelectric effect. For a historical discussion of Nobel Prizes with attention to Scandinavian perspectives, see Friedman (1989, 1990, 2011). I am very much obliged to Karl Grandin for his guiding me through the Nobel Archives.

attributes of the hagiographic genre in the history of science do not by themselves provide a sufficient explanation, but rather a post-factum rationalization. Even with a great scientist at hand and in the best of all circumstances, it is a rather improbable task to make a country as small as Denmark the world leader in any major field of science. Again, Scandinavians are more likely than outsiders to admit that "empire building on the periphery is no easy matter" (Friedman 1990, 193) and that individual virtues and accomplishments in science do not automatically translate into outsized institutional power and leadership. To illustrate the point to other readers, especially those who come from a large country, consider the earlier, but comparable case of physical chemistry.

Physical chemistry achieved recognition as a new university discipline at the close of the nineteenth century. One of its founders, Wilhelm Ostwald, was thoroughly self-conscious about his personal contribution to the field, writing about himself in the third person:

> In the history of modern science, it has become common to connect the name of Ostwald with the names of Van't Hoff and Arrhenius, although he did not make a comparable discovery at the same time. It is because the *organizational* factor is instilled in my personality, without which the new branch of science could not have been established so quickly and so widely. Through my appointment in Leipzig, the new discipline acquired its geographical and school-building focal point.[5]

Indeed, the other two founding fathers of physical chemistry contributed crucial ideas and discoveries but lacked institutional influence as they worked in small countries: Jacobus Van't Hoff in Holland and Svante Arrhenius in Sweden. A large university institute that functioned as the center of advanced teaching, pilgrimage, proselytizing, mass production of PhDs, and dissemination of the new science throughout the world was organized by Ostwald in Leipzig, Germany, a major power that could afford such a luxury in the 1890s. In a similar way, the institutional center of quantum theory normally would have developed in Germany—in all likelihood in Munich around Arnold Sommerfeld—had it not been for World War I.

The Great War left German science significantly weakened and internationally isolated, but it was still rather improbable for tiny Denmark to step into the role of a major center of scientific research. A closer inquiry is required to elucidate the factors that made such an unlikely outcome possible, at least in principle. We shall find out that, in addition to Bohr's great diplomatic skills and full-time commitment to institutional development, he relied on the very specific conditions in Europe in

---

[5]"Es ist in der Wissenschaftsgeschichte dieser Zeit üblich geworden, mit den Namen van't Hoff und Arrhenius, auch den Namen Wilhelm Ostwald zu verbinden, obwohl er nicht durch eine gleichwertige Entdeckung um dieselbe Zeit hervorgehoben wurde. Dies liegt daran, daß in meiner Person sich der *organisatorische* Faktor verkörperte, ohne welchen eine derart schnelle und weitreichende Gestaltung eines neuen Wissensgebietes nicht stattfinden kann. Denn die neue Wissenschaft gewann durch meine Berufung nach Leipzig einen geographischen und schulebildenden Mittlpunkt." Ostwald (1927, 2: 20). On the institutional development of physical chemistry, see Servos (1990) and Kormos Barkan (1992).

the 1920s.[6] In different historical situations, such as before or after the short interwar period, it would have been almost inconceivable for the world's main center of a discipline to function out of Copenhagen. Indeed, despite Bohr's ever increasing international reputation and best efforts, he was not able to maintain his institute's central position in theoretical physics after the end of World War II.

Denmark's small size meant that Bohr could only build his informal empire internationally, both in terms of its reach and the resources he drew upon. Instead of the otherwise mysterious "Copenhagen Spirit" as a metaphorical force that supposedly attracted young physicists from all over the world to his institute (Meyenn et al. 1985; Pais 1991), this study utilizes the concept of the network to document the growth of the Copenhagen-centered cohort of theoretical physicists. One does not have to define networks in a specifically Latourian way to be able to combine, in a non-discriminative manner, heterogeneous resources: people and monies, politics and institutions, ideas and gadgets, to mention only the most typical ones, without having to either reduce one element to another or oppose them as mutually exclusive types of causation (Latour 2005). In what follows, I will rely on this important feature of networks in order to reconstruct and analyze historically the diverse elements out of which Bohr constructed the international community of quantum physicists.

Another, older notion from the sociology of science—the center and the periphery—can also be applied here, but with a notable reservation (Ben-David 1971). For Copenhagen to acquire the reputation of the main node of quantum mechanics, an effective centralization of events and contributions did not have to be achieved in historical actuality, but rather in contemporaries' imagination. It was important for the situation to appear as such in the perception of many participants, subsequent commentators, and in the historical myths they created. The construction of physicists' disciplinary memory regarding the creation of quantum mechanics was also largely a product of the Copenhagen network and an important part of its success story. Some elements of this story, like any folklore memory, inevitably have to be revised by careful historical investigation.

Building an informal network often relies on oral rather than written communications and therefore cannot be recorded with sufficient detail in extant documents. My reconstruction combines information and hints from various sources, arranging them in spatial and chronological patterns. Most of the unpublished documents and oral history interviews come from three major collections: the Archive for the History of Quantum Physics (AHQP), the Niels Bohr Archive (NBA), and the Rockefeller Archive Center (RAC). One particular source deserves special mention: the annotated card catalog of Bohr's correspondence preserved in NBA, which lists letters in strict chronological order, rather than the alphabetical order of correspondents typically preferred by archives. This arrangement allowed me to follow Bohr's extraordinarily massive correspondence on a day-to-day basis, uncovering tacit relations between various aspects of his networking activities. Throughout the 1920s, Bohr dictated several hundred letters annually, an activity that must have taken no less of his time

---

[6]"The Institute... which had the foresight to grasp the unique combination of circumstances that were at hand some sixty years ago." Aage Bohr, "Foreword," in Robertson (1979, V).

and thought than his administrative chores as director and builder of the institute or his discussions with numerous visitors about quantum physics and life. Many of Bohr's letters used carefully constructed phrases and ambiguous diplomatic formulations, so much so, that their meanings and subtexts can often only be understood with the help of the typically more open and straightforward responses and reactions to them by Bohr's correspondents, who had access to additional kinds of information, including oral conversations and shared tacit knowledge.[7]

---

[7]A number of penetrating historical investigations have described and scrutinized the intellectual aspect of Bohr's research program and his contributions to quantum theory: (Heilbron and Kuhn 1969; Hendry 1984; Pais 1991; Darrigol 1992; Kragh 2012). The best account of the development of the Copenhagen Institute is in Robertson (1979), to which the present study is indebted for many clues and tacit hints. The later period of Bohr's institute during the 1930s is described thoroughly in Aaserud (1990).

Chapter 2
Scandinavian Settings

In July 1916, in the midst of the catastrophic European war, a thirty-year-old Dane returned from Manchester, where he had worked during the preceding two years, to his native Copenhagen to assume a professorship in theoretical physics. Upon landing, he reported back to his British mentor Ernest Rutherford that the boat had not been torpedoed and arrived safely.[1] By that time, Niels Bohr had already published what would later be judged his greatest contribution to physics, the quantum model of the atom (Bohr 1913). His theory allowed to calculate the spectra of the hydrogen atom with a method that, although not entirely explainable from the point of view of the usual mechanics and electrodynamics, was at least explicable as a set of rules based on the somewhat mysterious notion of the quantum. The future importance of the Bohr model was not yet obvious, even to Rutherford, and several more years would pass before it won wider acceptance among physicists.[2]

Rather than the esoteric mathematical theory, what probably meant more for Copenhagen officials as proof of Bohr's suitability for the professorship was Rutherford's offer of a position as reader in mathematical physics at the University of Manchester. Bohr's earlier attempts to obtain a professorship in Denmark had been unsuccessful. In 1912, the University of Copenhagen offered its only professorial position in physics to experimentalist Martin Knudsen. Since no second chair seemed forthcoming, and Knudsen was only a few years older, Bohr's prospects for solid academic employment appeared bleak.[3] The university's second, subordinate position of Docent in physics became vacant with Knudsen's promotion, and Bohr was appointed to it in September 1913. In March 1914, he submitted a proposal for the creation of a new professorship in theoretical physics, which was rejected by the

---

[1] Rutherford to Bohr, 31 July 1916.

[2] Bohr's classic papers of 1913 are reprinted and analyzed in (Aaserud and Heilbron 2013). For the history of Bohr's and other atomic models, see (Heilbron and Kuhn 1969; Heilbron 1977; Kragh 2012).

[3] "As conditions are in this country… it would be for many years, maybe forever, impossible… to get a scientific post at the university." Harald Bohr to C. W. Oseen, 7 March 1912. On Knudsen, see Pihl (1983, 393–396).

© The Author(s), under exclusive license to Springer Nature Switzerland AG 2020
A. Kojevnikov, *The Copenhagen Network*, SpringerBriefs in History of Science and Technology, https://doi.org/10.1007/978-3-030-59188-5_2

minister. A year and a half later, Bohr's second attempt, sent from Manchester where he was then on leave, finally succeeded.[4]

The University of Copenhagen thus obtained a second professorship in physics and with it also a specialty in theoretical physics, a relatively novel academic discipline which by that time had already established itself in the universities of neighboring Germany. Bohr's conditions, however, hardly matched those of his German peers. Valdemar Henriques, a university official and close friend of the Bohr family, dutifully warned him in advance:

> [I]n a cabinet meeting a couple of weeks ago, it was decided that the much desired civil servants salary bill will be introduced right after New Year. If the bill is passed, as everything indicates it will be, the act will call for a professorship in theoretical physics (or whatever it is called). This will probably mean that you will become professor next year... [I]f you become professor ordinarius here in Copenhagen it will most likely take some time before you can get even tolerably good working conditions – that may perhaps not happen before Prof. Prytz retires.[5]

Kristian Prytz taught physics at the *Polyteknisk Læreanstalt* (Higher School of Technology) which also maintained Copenhagen's only experimental laboratory facility for physics, sharing it with the university. Initially, Bohr had to be satisfied with one room in the *Læreanstalt* building and had neither a secretary nor a salaried assistant.

Denmark's neutrality during the war brought a boom for its international trade and general economy. University finances also improved so that in spring 1917, less than one year after Bohr's arrival, each of the two physics professors submitted formal proposals to establish a university institute in physics with an experimental laboratory. The obvious inspiration came from the institutionally influential model of research universities in the German Reich, where such specially built institutes had already become a norm in most branches of experimental science (Cahan 1985). They typically included a separate building with a large auditorium for lecture courses, classrooms for practical instruction in experimental skills, laboratory rooms in the basement for the professor's and his assistants' advanced research, and living quarters for the director's family. Bohr's proposal contained these usual elements, but also added some features that reflected Copenhagen's specific conditions. Later, through the Bohr institute's international fame, they acquired a more general, international, and trend-setting significance. He started with an idea of a reversed relationship between theory and experiment, to which he then added a locally important international agenda, and an orientation toward young researchers.

---

[4]Munch-Petersen (1925, 4: 283–287), Robertson (1979, 16–17); Bohr to Oseen, 11 August 1913, 10 March 1914; Bohr to Rutherford, 10 March 1914 (BCW 2: 557; 591–2), Knudsen to Bohr, 20 May 1915. On the crucial role of family connections in securing Bohr's professorial appointment, see Aaserud and Heilbron (2013).

[5]Henriques to Bohr, 23 December 1915 (BCW 2: 521–2).

2.1 Experiment Under Theory

Officially, Bohr's institute would be devoted to research in theoretical physics. Given such a strong disciplinary identity and subsequent fame, it may appear peculiar that in his initial plans and efforts, Bohr placed the main emphasis on establishing an experimental laboratory under his purview.[6] His early collaborator in experimental research, Hans Marius Hansen, studied spectroscopy in Göttingen in the same year when Bohr studied atomic physics in England, 1911/12. Both were back in Copenhagen in 1913, and it was Hansen who then reminded Bohr of the Balmer formula for hydrogen spectrum, which opened the way toward the completion of Bohr's atomic model and its connection to the wealth of spectroscopic data (Heilbron and Kuhn 1969, 264–265; Kragh 2012, 57). In 1913, Bohr served as an official opponent at the defense of Hansen's doctoral thesis in Copenhagen on the inverse Zeeman effect. Together, they performed experiments on the Stark effect, but without much success. While Bohr was on leave in Manchester during the war, Hansen replaced him as Docent in Copenhagen and was officially appointed to this post in 1918 after Bohr's promotion to professor.[7]

Bohr urgently needed to establish close contact with a qualified spectroscopist, so that he could check and confirm theoretical predictions from his theory. To answer his many questions and further hypotheses, he hoped to continue spectroscopic experiments with Hansen, but available facilities in Copenhagen could not even provide them with a single room for the laboratory. In his first proposal for the physics institute, Bohr insisted on the need for the professor of theoretical physics to supervise experimental research: "in order that theoretical investigations of this type can be conducted to advantage, it is necessary however that the scientists occupied in this way are given the opportunity to also undertake experimental investigations" (Robertson 1979, 21). He hinted at "several important foreign universities" that had similar arrangements, but did not offer a concrete example. An important precedent for such an institutional arrangement did exist at the Institute for Theoretical Physics in Munich, which opened in 1910. Its director, Arnold Sommerfeld, although a pure theoretical and mathematical physicist himself, had at his disposal a collection of instruments and was able to hire an experimental assistant and students to use this equipment for checking theories. This combination proved effective in a major discovery made at the institute, the diffraction of X-rays, proposed theoretically by Max von Laue and discovered experimentally by Walter Friedrich and Paul Knipping in 1912. The institutional power of theoretical physics to direct experimental

---

[6]Bohr to the Faculty of Science and Mathematics, 18 April 1917 (Robertson 1979, 20–22; Munch-Petersen 1923, 3: 316–329). See also references to a "small experimental laboratory" in Bohr to Rutherford, 9 December 1917; Bohr to Richardson, 15 August 1918.

[7]Robertson (1979, 16–17), Munch-Petersen (1925, 4: 305–307); Bohr to J. N. Brønsted, 14 June 1913; Bohr to Oseen, 3 March 1914; Hansen to Bohr, 23 September 1915.

research was rather untypical for German universities, where the opposite pattern—of the theoretical physicist working in a position institutionally subordinate to that of the professor of experimental physics—was traditionally dominant.[8]

The sprouting of physics into subdisciplines, along with the division of labor between experiment and theory, was not as natural and self-evident as it appears to us post-factum. Historically contingent, the disciplinary separation evolved in German universities in complex interrelation with matters of university curriculum, academic hierarchy, and control over scientific instruments. In the late eighteenth century, some enterprising professors of physics, such as Georg Christoph Lichtenberg, started advertising their classes as "experimental physics." At the time, the use of this term meant a promise that students would not only be listening to lectures, but also watching live experimental demonstrations, including exciting tricks with static electricity. Professors who utilized this innovative teaching style had to invest their own money into building or buying the necessary instruments, but they also were compensated by larger sums of student fees, as such entertaining courses quickly became more popular among students than boring lectures (Hund 1987; Hochadel 2008). On retirement, professors usually sold their private instrumental collections to their successors or to the university, and in the next generation, the practice of showing experimental demonstrations in class became a norm in most universities. Professors who proudly called their teaching discipline "experimental physics" not only used demonstrations during their lectures, but also controlled an expensive university collection of diverse physical instruments that came together with the job title.

By the mid-nineteenth century, when many German universities started hiring second teachers in physics, a division of curriculum needed to be established. Theoretically, one could imagine dividing the existing instrumental collections according to some conventional subject matter, say, mechanics, or optics, but in reality, the professor who was higher up in the hierarchy, i.e., the full professor, or *Ordinarius*, claimed and retained control over the entire collection of instruments and with it, the privilege to teach courses in experimental physics. Though often introductory or lower level, these courses had higher enrollment and were therefore more profitable to teach. The subordinate teacher (*Extraordinarius* or *Privatdozent*), lacking direct access to instruments, usually had to announce his lecture topic as mathematical or theoretical physics, which were often more advanced but attracted fewer students and resulted in lower pay. At the time, most physicists were expected to be able to perform both experiments and mathematical calculations, and although some may have preferred one activity over the other, the emerging disciplinary division within physics initially became pronounced in curriculum, and only later transformed into a research specialization. It correlated with the consecutive stages in a typical academic path: Many physicists in the second half of the century started their academic careers with junior appointments as teachers of theoretical physics and eventually advanced

---

[8]IMN 2 (281–285). On the discovery of X-rays, see Forman (1969). I am indebted to Paul Forman for pointing out to me the example of the Munich institute.

to the position of a fully established professor of experimental physics (IMN 1: 234–45; Hund 1987).

After 1870, during a new wave of institutional expansion when many German universities erected buildings for physical institutes, ordinary professors of experimental physics assumed positions of directors, while theoretical physicists usually held a subordinate position of a lower-level teacher or in-house theoretician, who assisted the experimentalist in his research. The established disciplinary hierarchy also explains why Jews disproportionately ended up in positions as theoretical rather than experimental physicists. Bohr was familiar with a similar institutional arrangement in Britain, where he served in Rutherford's Manchester laboratory in a subordinate position as reader in mathematical physics. Starting around 1890, some German universities succeeded in creating the second *Ordinarius* in physics, which needed a new disciplinary name, because the existing academic rule avoided having two full professors in one faculty with exactly the same job title. The status of theoretical physics thus gradually advanced from a mere division in teaching curriculum to a separate academic discipline allowing a full-level career, but this did not mean an immediate separation in actual scientific research. Most early occupants of the chairs of theoretical physics still worked in both theory and experiment, and in order to satisfy this as well as their rank of *Ordinarius*, some space was carved out of the existing buildings and some instrumental collections ascribed to the new institutes of theoretical physics, which were typically smaller than institutes and the collections placed at the disposal of experimental physics professors. The next generation of appointees circa 1900 already included some professionals, or pure theoretical physicists such as Sommerfeld, who inherited their institutions and instruments from less specialized predecessors but did not perform any experiments themselves.[9]

The status of the physics discipline in Denmark developed somewhat differently. Although German trends were followed in general, the University of Copenhagen established its chair of ordinary professor in theoretical physics before the institute in experimental physics. Several earlier proposals of such an institute were unsuccessful and, as of 1917, Copenhagen had only a small laboratory in the building of the *Polyteknisk Læreanstalt*. The need for a decent university institute was commonly recognized, and when a realistic possibility presented itself, both existing professors were already in officially equal positions to claim one. Knudsen proposed an enlargement of the existing laboratory into a full institute for experimental physics and physical chemistry. Bohr asked for a smaller institute for theoretical physics, but one that included an experimental laboratory.[10] Although not contradicting each other on paper, both documents basically implied the same—an institute in the German sense, with a lecture hall and a research laboratory—and competed for the same resources.

---

[9]Early directors of theoretical physics institutes, Ludwig Boltzmann in Munich and Vienna, Woldemar Voigt and Peter Debye in Göttingen, Theodor DesCoudres in Leipzig, also performed experimental research. Max Planck appears to be "the first pure theoretical physicist," and his institute in Berlin lacked a collection of experimental instruments (IMN 2: 33–54; Cahan 1985). Noteworthy, that in other university disciplines, such as chemistry or biology, emerging subdivisions separated the objects of study rather than the style of research.

[10]Bohr (1923), Munch-Petersen (1925, 4: 315–17); Bohr to S. H. Weber, 31 May 1917.

Indeed, only one proposal, Bohr's, would finally succeed, while Knudsen's was rejected. This bureaucratic outcome explains, in part, Knudsen's subsequent alienation from modern physics in general and quantum in particular, and his eventual shift to a research program in oceanography (Nielsen 1963, 24).

Although Bohr's status allowed him to submit an independent application for an institute, his job title made it more difficult for him to argue that an experimental laboratory should operate under his direction. His rhetoric included denials of disciplinary boundaries between experiment and theory:

> it is obvious that such a division conditioned by teaching considerations does not reflect a similar division of the scientific research in physics. And it must be especially emphasized that, considering the ways science has developed, fruitful theoretical research is totally impossible without concurrent experimental work which is required to test the correctness of the various possibilities presented by the theoretical assumptions.[11]

In order to justify upending the existing hierarchy of status and prestige between the two subdisciplines, Bohr invoked the archaic British term "natural philosophy," in which "philosophical thinking provides a basis for experimental investigations, since it is necessary to form an idea as to what questions to ask of nature in order to have a hope of obtaining fruitful answers. In recognition of this fact, all study of physics in England is called natural philosophy." He also employed linguistic novelties, specifying the institute's task as to "take up *experimental* investigations in certain new domains of *theoretical* physics."[12] It was not, however, this kind of rhetoric that explains why Bohr's proposal eventually won, but more mundane financial considerations.

In October 1917, the old-boy network of high-school graduates, mobilized by Bohr's classmate and businessman Aage Berlème, established a private committee with the task of raising money for the institute from individual and corporate sources. Denmark's most respected scholar, philosopher Harald Høffding, and adventurous entrepreneur Harald Plum joined the committee. By December 1917, Berlème had collected pledges for almost 80.000 kr. (about $20.000), which eventually sufficed to purchase the required piece of land. This investment certainly influenced the decision by the ministry and the parliament to approve the institute proposal in October 1918, just in time before the war's end and before Denmark's temporary financial prosperity gave way to a serious postwar economic and social crisis. Subsequent inflation and the rise in prices more than doubled the budgetary expenditures from the originally

---

[11] "Det er imidlertid indlysende, at en sådan af undervisningshensyn betinget deling ikke er udtryk for en tilsvarende spaltning af den videnskabelige fysiske forskning. Og det må da særlig fremhæves at frugtbar teoretisk forskning, således som videnskaben har udviklet sig er aldeles umulig uden samtidige experimentelle arbejder, der kræves for at prøve rigtigheden af de forskellige muligheder der frembyder sig for de teoretiske antagelser." Draft of Bohr's presentation to the University *Konsistorium*, 8 June 1917 (NBA).

[12] Bohr's speech at the dedication of the Institute for Theoretical Physics, 3 March 1921 (BCW 3: 293). Bohr to IEB, 27 June 1923 (RAC. Projects. Denmark).

approved sum of 200,000 kr. (+60,000 for the equipment), but the state commitment made in 1918 was not revoked.[13]

When Bohr's institute opened in early 1921, a new cultural precedent was set and later became known far outside Denmark. Though it would not become a usual practice for theoretical institutes to run in-house experimental facilities, allowing researchers to check hypotheses and conjectures at will, a somewhat different combination—that of an experimental laboratory under the direction of a professional theoretician who did not do experiments himself—would later function in many places, including J. Robert Oppenheimer's Los Alamos and Heisenberg's Max-Planck-Institut für Physik.

2.2 New Blood

Bohr's professorship did not entitle him to an assistant, whom he needed quite urgently because of his working style. From early on, Bohr experienced unique difficulties with writing. He was able to write letters, easily and masterfully, although later in his career he stopped handwriting them personally and typically dictated to his secretary, Miss Schultz. But when it came to scientific papers, Bohr was unable to write these alone, requiring a sparring partner for conversation, discussion, and dictation. Starting with his doctoral dissertation of 1912, he relied on the help of others to formulate his thoughts. Bohr's PhD thesis was written, in this way, by his mother, despite other relatives' protests. In 1913, his wife Margrethe wrote with Bohr his groundbreaking trilogy on the hydrogen atom and continued helping him write papers until their first son was born in November 1916. Thereafter, Bohr's new pupil and assistant, Hendrik Anthony Kramers, assumed the role of "helper."[14]

Kramers came from the Netherlands, where he had studied physics in Leiden with Paul Ehrenfest. He had passed his doctoral examinations but was discouraged from doing further research toward a PhD. In Ehrenfest's opinion, Kramers was talented but somewhat lazy. In August 1916, Kramers visited Copenhagen for a student meeting and approached local professors of physics and mathematics. Introducing himself in a letter to Bohr, he explained that he wanted to study abroad but in a country that was not involved in the war. Starting September 1916, Kramers became Bohr's private assistant and took notes on Bohr's lectures about problems in modern physics. The lectures were reportedly only attended by six or eight students, and, as reflected

---

[13]Berlème to Knud Faber (university rector), 19 October 1917; Berlème to Bohr, 3 May 1918; Bohr to Berlème, 13 September 1918 (NBA). For a detailed account of the fundraising and the construction of the institute, see Robertson (1979, 23–38). During the 1920s, the Danish krone was worth about 20 cents US, albeit subject to large fluctuations due to inflation.

[14]Aaserud and Heilbron (2013); Interview with Fru Bohr, 23 January 1963, 14–15 (AHQP). Bohr had a secretary from the early 1920s onwards (BCW 3: 24).

in Kramers's notes, they already included material on quantum theory and atomic models.[15]

Private assistants were paid with soft money obtained by professors from various outside sources. Denmark had a system of organized private philanthropic support for research unusually well-developed for that time. As early as 1871, the largest Danish brewing company Carlsberg established a laboratory for research on fermentation, and in 1886 added a special foundation (as a gift to the Royal Academy of Arts and Letters) to maintain the regular expenses of the Carlsberg laboratory and to give out surplus funds for the promotion of science in general. The Carlsberg Foundation's board of directors consisted of five members of the academy and distributed small annual grants to almost every member of the academy and to the faculty of the University of Copenhagen.[16]

In 1911/1913, the Carlsberg Foundation supported Bohr's studies in England and his research on atomic models, receiving as part of the report a reprint of the famous paper "On the Constitution of Atoms and Molecules." In 1913/1916, the foundation gave out a grant for Bohr's and Hansen's attempts to perform spectroscopic experiments. From 1916 on, it provided Bohr with an annual sum "for assistance in computations," which initially sufficed to pay a salary for one assistant (2,000 kr. in 1916; 3,500 in 1917–19). In later years, the amount increased to 12,000 and then to 18,000 kr. (or about $3,000) annually and contributed to partial support of five to six students and assistants, most of them Danish or Scandinavian.[17] Bohr also received some additional grants from Carlsberg for scientific equipment. His contacts with Danish foundations provided Bohr with an important experience and prepared him for further dealings with international philanthropic organizations.

Kramers took over from Bohr's wife the role of an assistant in writing, but he needed a career of his own and was professionally qualified to do more than just "putting phrases together… giving them a good form."[18] He soon became engaged in mathematical calculations and in close scientific collaboration and intensive discussions with Bohr. Within a year, Kramers produced an important result of his own—the calculation of the Stark effect in hydrogen, the splitting of spectral lines in an external electric field—and an inevitable question of the division of labor and credit arose, causing a somewhat uncomfortable exchange of written letters between the two.

> [O]ne may expect to obtain a large number of significant results in the nearest future; hence, it is absolutely necessary that both of us are perfectly clear about the form of our collaboration, and that this be arranged in a way that is reasonable and justifiable for both of us… [I]n talking

[15]Nielsen (1963, 23); Kramers to Bohr, 25 August 1916 (BCW 2: 537); Ehrenfest to Bohr, 10 May 1918; Kramers's Notebooks: September 1916. *Atoommodellen* (AHQP). For a biography of Kramers, see Dresden (1987).

[16]Glamann (1976, 2002) and Danes proudly claim that the Carlsberg Foundation was established earlier than similar foundations in other countries: Carl Zeiss in Germany (1889); Nobel in Sweden (1900); Rockefeller in the USA (1913); Welcome Trust in Britain (1936); van Leer in the Netherlands (1949).

[17]Carlsbergfondet (1930); The list of Carlsberg fellows at Bohr's institute includes 35 names for the period 1916–1935 (NBA).

[18]Interview with Fru Bohr, 30 January 1963, 12 (AHQP).

with you last night about the calculations, I got the feeling that you perhaps do not think that the manner of continuing our collaboration that we talked yesterday morning – temporarily to apply jointly the new theoretical viewpoints to special problems and discussing together the results of the calculations—is the wisest for you, but that you might prefer to try, more independently, to work out some problem,[19]

suggested Bohr. Kramers was confused, but it was for him to propose a solution:

If I only knew well and clearly what the matter is. The whole thing appears to me as some vague half ethical and half practical question, and I was therefore so glad the morning of the day before yesterday when you said that it wasn't anything serious at all. But now you write me a letter and ask if I wouldn't prefer to work more independently... Just as little as I can refrain from working a little independently, just as little can I refrain from later asking your advice; for you can further the matter so very much by the philosophy that you can put into it... Or does the whole thing mean only that we must accurately, and for each special subject, agree upon what you publish alone, what we publish jointly, and what I publish alone. Nothing is easier than to do that.[20]

Eventually, they worked out a mutually satisfactory arrangement. Bohr's paper, "On the Quantum Theory of Line Spectra," written with Kramers's help in 1918, developed the general "correspondence argument," or the use of analogy with Fourier spectra of classical radiation to predict the properties (intensities and selection rules) of quantum transitions. Kramers's own paper of 1919 furnished Bohr's reasoning with sophisticated mathematical calculations of the Stark effect (Bohr 1918; Kramers 1919).[21] Research accomplished along these lines provided Kramers with his doctoral thesis, which he defended in Leiden, with Ehrenfest, in the spring of 1919, after which he received a promotion in Copenhagen to *Eneassistent*, or salaried university appointee.[22] His collaboration with Bohr continued along the same pattern of division of labor throughout his entire tenure in Copenhagen until the end of 1925. Bohr usually published a more general paper, referring in advance to mathematical calculations and formulae obtained and subsequently published separately by Kramers, while Kramers in his paper referred to and thanked Bohr for directions. Despite a warning from a Dutch friend—"you cannot stay Bohr's appendix forever"—Kramers was confident that the collaboration was mutually beneficial:

---

[19] Bohr to Kramers, Hellerup, 15 November 1917.

[20] Kramers to Bohr, Hellerup, 16 November 1917.

[21] Their point of departure was the famous introduction of transition probabilities for spontaneous and induced emission of radiation in Einstein (1916). Bohr reinterpreted these results away from the concept of light quanta, implied by Einstein, toward the analogy with classical electromagnetic radiation. For the clearest analysis of the correspondence principle, see Darrigol (1992, 121–149). Most historical descriptions assume that Bohr's general argument preceded and logically led to Kramers's calculations, but this sequence simply reflects the order of the eventual publications, not necessarily the genesis of ideas.

[22] Bohr (1923); Bohr to Ehrenfest, 25 January 1919.

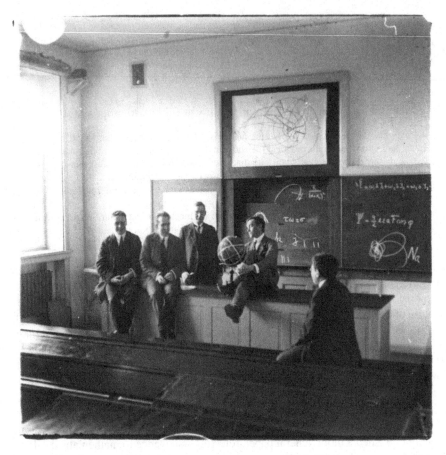

Fig. 2.1 Hansen, Bohr, Kramers, Rosseland, and Ehrenfest in 1921 (NBA)

one thing is certain: never shall I be able to fathom what precisely were the struggle and the victories in the life of my best friend. His line of thought, his sensing and understanding of the world I shall never come to know. Here the intellect can accomplish nothing, the instinct a little, love alone everything (Fig. 2.1).[23]

In March 1917, Kramers traveled to Sweden to give a lecture in Stockholm on his latest investigations with Bohr. At the time, neither Sweden nor Norway had established chairs in "theoretical physics," but only in somewhat old-fashioned "mathematical physics," the name that usually correlated with their occupants doing research on the problems of classical physics. Although Kramers did not succeed in overcoming Swedish physicists' skepticism toward quantum theory, he did start a friendship with Oskar Klein, who was then an *amanuensis* at Arrhenius' Nobel Institute for Physical Chemistry. In 1918, Klein received a fellowship to study in

---

[23]Romein to Kramers, 20 February 1923; Kramers to Romein, 28 October 1924, quoted in Radder (1982, 235–237).

Germany, but with the war still going on, he decided to use it to come to Copenhagen for the summer of 1918 and then again for the second half of 1919. When Kramers fell seriously ill with typhoid, Klein replaced him as Bohr's assistant and received part of his salary.[24]

Both Klein and the Norwegian Svein Rosseland from Oslo shared a general theoretical interest in science and mathematical abilities, but were not well suited for more practical, experimental work with their supervisors in home countries. This is how Rosseland's mentor Vilhelm Bjerknes and Olof Arrhenius, Svante's son, recommended him to Bohr. Rosseland failed to obtain a Norwegian fellowship, but Bohr secured some of the Carlsberg funds to bring him to Copenhagen in the fall of 1920, during which time Rosseland and Klein authored an important paper on atomic collisions.[25] Supported by Bohr with either Carlsberg or Rask-Ørsted grants, they both spent at least half of each of the subsequent years in Copenhagen: Klein until 1922, Rosseland until 1924. Like Kramers, they defended doctoral theses at home universities: Klein in Stockholm in spring 1921 on his Swedish work on electrolytes, Rosseland in Oslo in 1923 on astrophysics, the subject of his personal interest. Opportunities for getting a decent position anywhere in Scandinavia were still very rare. After spending a year as docent in theoretical physics in Lund, Klein, with the help of Bohr's recommendation letters, went overseas to become in September 1923 an assistant professor at the University of Michigan, Ann Arbor. Rosseland departed for the USA as well in the fall of 1924, with a postdoctoral fellowship from Rockefeller's International Education Board.[26]

The initial example set by Kramers established the characteristic pattern of the division of labor at the Copenhagen institute. Bohr led the oral discussion with visiting fellows regarding general physical and philosophical aspects of research, while *medhjælper* (Kramers and later other assistants) performed calculations and wrote more technical papers for themselves and also general ones for the professor, following his dictation. The process of writing with Bohr was meticulous and often frustrating. "Bohr usually worked on two or three things at a time. He would work with one of the younger physicists in the morning, with another in the afternoon, and with a third in the evening. In spite of his great kindness, he demanded much of his collaborators, and they would be exhausted at the end of the day," recalled Nielsen on his experience of working through nine consecutive proofs of Bohr's 1931 lecture "Light and Life" (Nielsen 1963, 24). Always striving for correct formulations and never satisfied with the existing ones, Bohr's quest for subtlety and precision often resulted in the opposite—complexity and obscurity of meaning—though the latter

---

[24]On retroactive, rhetorical construction of the concept of classical physics, see Staley (2005). Kramers to Bohr, 17 March 1917; 27 January 1919, 26 September 1919; Klein to Bohr, 27 March 1918.

[25]Bjerknes to Bohr, 27 January 1920; Rosseland to Bohr, 6 August 1920; Bohr to Bjerknes, 12 January 1921; (Klein and Rosseland 1921).

[26]Klein to Bohr, 25 October 1922; Rosseland to Bohr, 18 September 1924. Two Danish students of physics—Fricke and Rud Nielsen—who also assisted Bohr occasionally, also had to leave for the USA.

usually did not diminish the authority of Bohr's voice and how his publications were received.[27]

The initial success of his collaboration with Kramers also prompted Bohr's expectation—a somewhat peculiar one for a professor who was himself only 35 years old—that new ideas in quantum physics would be coming from the younger generation:

> This leads me to emphasize once more that this Institute is not intended solely for scientific research but also to be a homestead for the teaching of physicists and others with special interest in physics… In fact, it is in the nature of scientific researches that no one dares give definite promises for the future; we must be prepared that, on the road that at this moment is believed to lie open and smooth before us, obstacles can pile up which entirely bar the road, or to the overcoming of which entirely new ideas are required. It is therefore of the greatest significance not just to depend on the abilities and powers of a limited circle of researchers; but the task of having to introduce a constantly renewed number of younger people into the results and methods of science contributes in the highest degree to continually taking up questions for discussion from the new sides; and, not least from the contributions of the younger people themselves, new blood and new ideas are constantly introduced into the work (BCW 3, 283–301).

This quote from Bohr's speech at the dedication of the Institute for Theoretical Physics on 3 March 1921 may sound like an astonishing prophecy—made true four years later by the crisis of Bohr's "old quantum theory" and the birth of new quantum mechanics—in almost every respect but one. His institute never developed a strong focus on teaching. Bohr withdrew from lecturing to students in 1920, when Kramers took over his teaching duties as his salaried assistant, while also continuing to help Bohr write papers. In general, little attention was paid to educating Danish physicists: In the 20 years between the two world wars, only five doctoral dissertations were defended at the institute, four in experimental physics and only one theoretical.[28] This is understandable, when one recalls that there were hardly any academic positions for physics PhDs in Denmark. Unlike prototypical German institutes with their large numbers of doctoral students, Bohr's institute never became a degree-granting factory in which research was done mainly by students working toward their doctorates. Instead, it hosted young physicists who had received such degrees elsewhere—postdoctoral researchers in modern terminology—and advanced their careers internationally, since the small Danish academic system allowed little room to expand nationally.

---

[27] Ebbe Rasmussen, "Medhjælper hos Niels Bohr" (1955, NBA). Heisenberg, who later served as one of the 'helpers,' left the following description: "Bohr would always change the sentences again and again. He could have filled half a page with a few sentences and then everything was crossed out and changed again. And even when the whole paper was almost finished… the next day everything would be changed over again… The final text of Bohr's paper was so subtle and he would think about half an hour whether in a certain case he would use indikativ or the konjunktiv and so on" (Kragh 2012, 193).

[28] Kramers's 1920–1922 lecture drafts, 1923–1924 seminar plans in his notebooks (AHQP); Dissertations defended at Bohr's Institute: Sven Werner (1927), J. C. Jacobsen (1928), Ebbe Rasmussen (1932), Chr. Møller (1932), R.E.H. Rasmussen (1936) (NBA).

2.3 Meanings of Internationalism

Internationalism and nationalism do not exclude each other, only internationalism and isolationism do. In many historical cases, international activity is driven by national ambitions; it is not accidental, for example, that the rise of nationalism in late nineteenth-century Europe coincided with the heyday of such events as World Exhibitions, supranational scientific organizations, Olympic games in sport, industrial and economic exchange across the borders. World War I destroyed that system of competition between nations and produced a wider variety of internationalisms and of their combinations with different national agendas.[29]

The German prewar style of scientific internationalism (providing convenient places for study and research at their universities to foreign students and plentiful space in German scientific journals to foreign authors) was criticized, in particular by the French and the Belgians, as cultural imperialism. Of course, the system had helped positively reinforce the advanced position of German science versus its European rivals, and German professors and officials were certainly aware of that. After the war, spokesmen for science on the side of the victorious Entente expressed determination to put an end to German language domination in the sciences and, driven by their own nationalistic feelings, established a new version of international order based on the exclusionist principle. Under the pretext of avoiding cultural imperialism, they did not allow Germany and its wartime allies to join the newly organized international scientific organizations, such as the International Research Council, or to participate at international conferences and meetings. The scientific boycott lasted almost eight years, until 1926, during which time Germany had to redirect its main international activities in science toward neutral countries, Soviet Russia, and Japan (Kevles 1971a; Forman 1973; Schröeder-Gudehus 1978, Cock 1983; Johnson 2017).

In the postwar situation of "hostile political camps," neutral Scandinavian countries saw an opportunity to develop international scientific agendas of their own. Their new policies were first indicated already in June 1917 at a meeting in Kristiania (Oslo), held on the initiative of Fredrik Stang. Representatives of political and academic circles from Norway, Sweden, and Denmark met there to discuss strategies for the postwar period. In the midst of the war, they were able to predict quite correctly that hostilities between scientists from major countries would not cease immediately at the end of the armed conflict, concluding that it would become the neutral countries' obligation, as well as their advantage, to initiate the postwar restoration of international contacts in science (Munch-Petersen 1925, 4: 50; Knudsen and Nielsen 2012). Although representatives from the three Scandinavian countries understood the situation quite similarly, their own local rivalries prevented them from cooperating in any joint course of action. As soon as the meeting was over and participants returned home, each country started developing somewhat similar but separate—and competing—international activities. After the war ended, each had internal debates

---

[29]Forman (1973), Somsen (2008), Walker (2012). For a history based on the simple opposition between nationalism and internationalism in science, provisionally defined as "involving scientists from three nations or more," see Crawford (1992, 38).

on whether to side with victorious allies in the new version of the international order in science. Typically, neutral countries agreed to join the new international scientific bodies, but negotiated the privilege to maintain separate contacts with German colleagues (Desser 1991; Lettevall et al. 2012).

The chief spokesman for science policy in Sweden, Svante Arrhenius, argued in 1919 that the war had given Sweden a new role in international science: as a mediator between the hostile big countries that "for a long time will be overpowered by a decided repugnancy against coming in close connection with one another." The impossibility of direct contacts between larger adversaries raised the importance of indirect ones, through neutral channels. If international research institutes were to be established in neutral countries, they would be able to benefit enormously from the existing situation: "A great opportunity is now open to the countries that were neutral during the war, hence also to the Scandinavian countries, to successfully apply the method which Germany applied with such eminent results, and which consisted of inviting the whole world to their scientific institutions." (Arrhenius, quoted in Widmalm 1995, 346.)

Arrhenius knew firsthand the advantages that the old international system had provided for German science, as he had worked in Ostwald's institute in Leipzig for a time before the war. It turned out, however, that Sweden's most successful mediating function would be performed not by an internationally oriented research institute, but through the careful evenhanded diplomacy of Nobel Prizes. The Danes acted with lesser publicity, but on 4 October 1917, the Ministry of Education appointed a committee of a dozen scholars and politicians to make recommendations regarding Denmark's postwar role in international scientific affairs. The committee, headed by Peter Munch, professor of history and the minister of defense at the same time, which reflected its perceived importance, understood that

> [International] undertakings … which were situated in the big countries before the war will not be able to continue… In the future, due to [the Scandinavian countries'] neutrality during the war, they will be able to achieve an influence that is disproportionate to their population. The initiative for international cooperation within science must essentially originate from them; many of the threads that will again tie the fighting [sides] together will be in their hands for a long time.[30]

As a practical suggestion, the committee proposed establishing a special foundation, subsequently named the *Rask-Ørsted Fond*, with the principal aim to "support the international scientific cooperation and secure the position of our country in it." The foundation was supposed to complement the existing, internally oriented Carlsberg foundation by

> taking up the international scientific responsibilities, as far as its resources allow it, by organizing headquarters for them in Denmark, establishing connections to the right figures,

---

[30]"De vil i Fremtiden i Kraft af deres Neutralitet under Krigen kunne faa en Indflydelse, der ikke staar i Forhold til deres Folketal. Initiativet til internationalt Samarbejde ogsaa indenfor Videnskaben maa væsentlig udgaa fra dem, mange af de Traade, der paany skal knytte de stridende sammen, vil i lang Tid komme til at ligge i deres Haand." "Oprettelse af Rask-Ørsted Fondet" in Munch-Petersen (1925, 4: 50–55, on 51).

both Danes and foreigners, establishing the right connections to all sides, undertaking the printing and publication in this country of their results. ...Further it must be emphasized that this case is of greatest national significance. Our country will hereby get the opportunity for an advanced position; through undertakings that are in this way centralized in Denmark, many Danish scientists will be stimulated and able to make use of their skills. Through the production and publishing of international publications, Denmark will be noticed and will have the opportunity to show its capacities, just as Danish science will be able to avoid losing its national character by being dispersed in foreign journals.[31]

This Danish strategy of scientific internationalism is referenced in the statement by the rector of the University of Copenhagen, who was also one of the members of the Munch committee, at the opening ceremony of Bohr's institute in March 1921: "We are proud of you and we expect much of your work. You have succeeded in gathering around you both Danish and foreign scientists and had thereby in the finest way continued the international collaboration that was broken off by the World War." (Robertson 1979, 9). Bohr formulated the goals of his institute accordingly, "to receive the properly qualified foreign physicists who wish to work there and to offer them suitable working conditions."[32] Subsequent developments showed that the program that Arrhenius had envisioned for postwar Sweden was actually realized most successfully by Bohr in Copenhagen. The two countries' activities eventually proved that they were not only competing along a similar track, but also mutually supportive vis-à-vis others. Swedish historian Sven Widmalm concludes: "The success of new fields like quantum physics, cultivated in neutral Denmark and marked by scientific honours from neutral Sweden, no doubt helped to preserve the status of the Nobel Prize." (Widmalm 1995, 360).

---

[31]"Fondets Formaal vil i første Række være at tage de internationale videnskabelige Opgaver op, saavidt dets Midler tillader det, ved at organisere Hovedsæde for dem i Danmark, knytte de rette Kræfter til dem, baade danske og Udlændinge, etablere de rette Forbindelser til alle Sider, paatage sig Trykning og Udgivelse her i Landet af deres Resultater... Det maa endvidere fremhæves, at denne Sag har en stor national Betydning. Vort Land faar herved Mulighed for en fremskudt Stilling, mange danske Videnskabsmænd vil gennem Foretagender, der saaledes centraliseres i Danmark, blive rigt befrugtede og kan faa Anvendelse for deres Kræfter. Danmark vil ogsaa ved internationale Publikationers Fremstilling og Udgivelse blive bemærket og faa Lejlighed til at vise, hvad det formaar, ligesom dansk Videnskab vil kunne undgaa at tabe i national Præg ved at spredes ud i udenlandske Tidsskrifter." (Ibid., 52). The money for the internationalism of the *Rask-Ørsted Fond* came from the selling of a colonial possession, the Danish West Indies, to the United States in 1916 (Knudsen and Nielsen 2012, 118).

[32]Bohr to IEB, 27 June 1923 (RAC. Projects. Denmark).

Chapter 3
International Networking

3.1 The Manchester Link

When Bohr became a professor in Copenhagen, he had some connections in Sweden, few in Germany, and almost none in France. Most of his foreign contacts were in Britain, in whose scientific community he was well socialized. This did not mean, however, that his atomic theory was accepted among British colleagues. Even Rutherford had reservations. The Bohr theory of 1913 could then be viewed as an attempt to hybridize visual atomic models, a characteristically British pastime, with the quantum approach from Germany. Rutherford was naturally skeptical about the latter: "Your ideas as to the mode of origin of spectrum and hydrogen are very ingenious and seem to work out well; but the mixture of Planck's ideas with the old mechanics make it very difficult to form a physical idea of what is the basis of it."[1] Even years later, Rutherford remained sympathetic but uncommitted to Bohr's theory. He claimed— almost in Ian Hacking's sense—to be able to visualize electrons and alpha particles, but not the mysterious quanta (Birks 1963, 39). A similar, at best reserved attitude toward quantum theory was shared by most established British professors at the time. The physicists who would start using it emerged from the younger cohort, who wanted to think differently.[2]

The war created a major crack between generations, including academic generations. What divided scientists most was not so much their physical age, but the experiences of war. The older group consisted of those who were lucky enough to take up their professorships before August 1914 and were thus mostly able to stay active in academic profession and research throughout the war years. Members of the younger generation had their scholarly progress interrupted either by participation in actual combat or by some war-related service. Resuming their academic careers

---

[1]Rutherford to Bohr, 20 March 1913 (BCW 2: 583). On the reception of Bohr's theory, see Kragh (2011, 2012, Chap. 3).

[2]"There were so many political things that then happened in the world and about which people have different views and young people sometimes wanted to think separately from older people." Oral History interview with the Dutch physicist J. M. Burgers (AHQP).

© The Author(s), under exclusive license to Springer Nature Switzerland AG 2020
A. Kojevnikov, *The Copenhagen Network*, SpringerBriefs in History of Science
and Technology, https://doi.org/10.1007/978-3-030-59188-5_3

after 1918, they sought to compensate for the lost four years and were more inclined to take up new paths and adopt radically new research strategies. Charles Galton Darwin (Darwin's grandson) was one of those younger physicists trying "to catch up with 4 years of science." After having served in the sound ranging of airplanes in France, he left the army in February 1919 and sent a letter to Bohr, resuming their relationship and exchanging scientific ideas. Darwin felt that "the fundamental basis of physics is in a desperate state. The great positive successes of the quantum theory have accentuated... also the essential contradiction on which it rests." As a remedy, he proposed

> to knock away the props of classical physics one by one and find, after a particular one has been removed, that our difficulties have become reconciled. It may be that it will prove necessary to make fundamental changes in our ideas of time and space or to abandon the conservation of matter and electricity or even as a last forlorn hope to endow electrons with free will.

Personally, Darwin preferred to think that "contradictions in physics all rest on the exact conservation of energy" and his preferred solution thus involved "denying that conservation is anything more than statistical." The language of electrons' "free will" was apparently too British for Bohr, but the draft of his reply admitted readiness "to take the most radical *or rather mystical* views imaginable" with regard to the daunting problem of the quantum interaction between matter and radiation. Five years later, Bohr would openly express support for the acausal, statistical conservation of energy, but in 1919 he was not yet prepared to declare which, if any, of Darwin's hypotheses he could endorse.[3]

They had worked together in Rutherford's laboratory, and when the war started in 1914, Bohr became Darwin's temporary replacement in the position of reader in mathematical physics. As a neutral national, Bohr was spared from war service and could continue his research, which allowed him to rise to a full professorship before the war's end. In the eyes of Darwin and others who returned to science, this demonstrated what they, too, could have achieved, had not their careers been interrupted. Established physicists in Britain were still mostly engaged with prewar lines of research. Quantum theory would become represented about five years later, with the promotions of Darwin and Ralph H. Fowler to professorial ranks. In his letter to Bohr, Darwin complained about the conservative attitudes of senior colleagues who, like J. J. [Thomson],

> seem to disregard every thing that has been done since about 1900... I am doing my inade-quate best to talk to people about quanta; everybody accepts them here now (which is better than it was in 1914 at any rate), but I don't think most of them realize their fundamental importance or have studied the arguments in connection with them. Rutherford said he was trying to persuade you to come here next term. I do hope you will. There are plenty of very intelligent people, only under the blighting influence of studying such things as strains in

---

[3]Darwin to Bohr, 20 July 1919, with an enclosed manuscript "A Critique of the Foundations of Physics"; Bohr to Darwin, July 1919, draft of a presumably unsent letter; "or rather mystical" is inserted into the sentence above the line.

the aether, they none of them know what is worth doing. If only they could have the right direction pointed out, I think a lot of good work could be done here.[4]

Rather than working in Britain himself, Bohr was more interested in bringing British scientists to Copenhagen. During his short visit to Manchester in spring 1919, he specifically looked for an experimental physicist capable of conducting spectroscopic research and of checking his theoretical hypotheses about the atom. He first turned to G. A. Hemsalech, a Pole, whose private means allowed him to conduct professional quality research in spectroscopy. Because of the war, Hemsalech was forced to abandon his personal laboratory in Paris and moved to Manchester. Bohr wrote to him in October 1919, once he learned about the approval of the Rask-Ørsted Fond: "Today… a bill is passed in the parliament about the establishment of an institution for < procuring > facilities for scientists from other countries to come and work at the Lab in Copenhagen in order to promote the international scientific intercourse and I am sure that in a few months I shall be able to… [make] you an offer which should make it possible for you to live here without expenses." In December 1919, Bohr repeated the invitation, but Hemsalech politely declined. He was hoping to return to his Paris laboratory, where experimental installations remained intact and feared that a move to a new and not yet equipped facility would prove detrimental to his work.[5]

Bohr's next choice, H. F. Biggs, also a former Manchester acquaintance, spent his war years sound ranging. In 1920, he held a minor position in Oxford, complaining, just like Darwin, about the situation there:

> it is not altogether satisfactory as the lab is so poor (in equipment, technical skill such as glass blower, and money) and Townsend is so distrustful of new theories. I don't think he has read anything about the quantum theory, and nothing more about relativity than Einstein's popular book. He is always wanting me to do research on the discharge between a wire and a cylinder and that sort of things. Rutherford was bad enough in trying to prevent men following up what they were interested in, but Townsend is ten times worse.

For Biggs, however, these difficulties did not outweigh the obvious advantages of Oxford, and, with the hope of receiving a possible research grant, he became far less enthusiastic about coming to Denmark:

> I should like nothing better than to go and work in Copenhagen for a bit. I might even, some day, be able to get off for a term… In any case whether I ever can go or not, I hope your schemes for getting people over to work in a lab of your own will have great success. Here, I am sorry to say, Lindemann as well as Townsend is rather indifferent about your theory, though I believe Merton is being converted.[6]

To convince a scientist from Britain to work in Denmark proved difficult: A temporary move to a peripheral location offered no advantages for their careers

---

[4]Darwin to Bohr, 30 May 1919. In his talk "La structure de l'atome" at the Solvay conference in 1921, Thomson did not mention Bohr's model.

[5]Bohr to Hemsalech, [August–October] 1919; Hemsalech to Bohr, 2 January 1920. Bohr also invited W. Makower (Bohr to Makower, 21 February 1920).

[6]Biggs to Bohr, 8 July 1918; 3 February 1919; 16 September 1920; Bohr to Biggs, [January] 1921.

back home. Eventually, Bohr's Manchester connections bore fruit, but not in Britain, and not in spectroscopy. The Hungarian George von Hevesy studied radiochemistry and knew Bohr from 1912, when they both worked in Rutherford's laboratory. He was also one of the earliest supporters of Bohr's atomic theory. During the war, Hevesy served in the Austrian army and in early 1919 was appointed as a professor and director of the physical laboratory at Budapest University. That summer, he traveled to Berlin, Munich, and Copenhagen, where he met with Bohr. Meanwhile, the short-lived Hungarian Bolshevik government was ousted by counterrevolutionary forces and replaced by an ultraconservative, anti-Semitic regime. Upon his return to Budapest, Hevesy suffered political attacks and lost his post. In October 1919, he informed Bohr of his desire to work in Copenhagen in the spring. While the institute was still under construction, Bohr arranged for Hevesy to work temporarily in the physicochemical laboratory of Johannes Brønsted. With continuing support from the Rask-Ørsted Fond, Hevesy lived and worked in Copenhagen until 1926.[7]

As a radiochemist, Hevesy could not satisfy Bohr's most urgent need to conduct experiments in optical spectroscopy. This job fell to Hansen, who was assisted by the Danish student J. C. Jacobsen (later also by Sven Werner) and Japanese visitors Toshio Takamine and later Yoshio Nishina. Problems with setting up the large grating spectrograph purchased with Hemsalech's help in England caused some delays. Eventually, Hansen proved more successful as a scientific entrepreneur, fundraiser, and administrator than as a spectroscopist. In 1923, upon becoming the university's third professor in physics, he left the institute to start his own program in the medical applications of X-rays and biophysics. Experimental spectroscopy never managed to develop in the Copenhagen institute to a degree that could fully satisfy Bohr's initial hopes and emphasis placed on it, and it would remain in the shadow of other, more successful activities at his institute.[8]

3.2   First German Connection

Initially, little pointed to the future strong German presence at the institute. Most of Bohr's early contacts in Germany were through his younger brother Harald, a mathematician with close ties to the Göttingen mathematical circle. Bohr had visited Germany only once, in July 1914, accompanying his brother to Göttingen and Munich, whereby he gave talks at Debye's and Sommerfeld's colloquia. The outbreak of the war forced the brothers to cut short their trip and planned vacations in Tyrol.[9] During the war years, Bohr in Britain exchanged a few letters with German physicists, particularly Sommerfeld, but no direct postal communication was possible between the warring countries. Apprehensive of the military censors, Bohr also could

---

[7]Levi (1985); Hevesy to Bohr, 25 October 1919; Bohr to Hevesy, 30 November 1919; 1 June 1920.
[8]Robertson (1979, 44–47); on Hansen, see Pihl (1983, 399–400). For Bohr's 1921 expectations that the "spectroscopic research will occupy first place" in his institute, see BCW 3 (296).
[9]Bohr to Hevesy, 1 July 1914; 8 August 1914; Bohr to Oseen, 28 September 1914.

not receive or send letters written in German. His letters written in English had to travel from Manchester to the neutral Copenhagen, where they were translated into German by either Harald or by their mother and mailed to the German addressee. The replies traveled back via the same indirect route and translation methods.

In Germany, Bohr's 1913 theory of the atom and the hydrogen spectrum found earlier and noticeably greater interest than in Britain, though the initial reaction was also mixed. Sommerfeld's response mirrored that of Rutherford almost exactly. He had no problems accepting quanta (originally a German concept), but the British taste for visualized atomic models was still foreign to him: "The problem of expressing the Rydberg-Ritz constant by Planck's h has been on my mind for a long time... Though for the present I am still rather skeptical about atomic models in general, calculating this constant is undoubtedly a great feat."[10] But Sommerfeld's attitude toward Bohr's theory improved rather quickly in 1915, once he found a way to advance it further, into what would become known as the Bohr–Sommerfeld quantum theory of multiple-periodic systems. Sommerfeld's generalized quantum condition allowed him to calculate elliptical orbits of electrons within the atom and more complicated spectroscopic effects. He dropped his skepticism regarding the atomic models, too, and in 1918 even advised the *Deutsches Museum* in Munich on how to prepare exhibits with models of atoms and molecules according to the newest theoretical ideas. Sommerfeld's support and improvement of Bohr's theory of the atom paved the way toward its wider acceptance in Germany and contributed to Bohr's growing reputation there. But the Sommerfeld school also presented a challenge for Bohr, since results obtained in Munich by the war's end, in many aspects, surpassed Bohr's own progress in developing his theory.[11]

Sommerfeld's most important advantage was his access to active collaborators and assistants. He had much closer ties with experimental spectroscopists, especially Friedrich Paschen at Tübingen University, with whom he was constantly exchanging ideas, results, and information on the new measurements of atomic spectra. Moreover, Sommerfeld could put to work on mathematical calculations a string of talented pupils. From 1906 on, he ran a seminar on theoretical physics at the University of Munich, which was not just a colloquium but a "school" in the German sense, i.e., a subsidized institution for the training of advanced teachers, PhDs, and future academics—functionally analogous, if less formal, to what came to be called "graduate school" in North America. During the war, when many of his German students were serving in the military, Sommerfeld was assisted by Paul S. Epstein, who was interned in Germany as a Russian subject, and by Adalbert Rubinowicz, a Pole. Epstein applied the Bohr–Sommerfeld methods to the Stark effect, or the splitting

---

[10]"Das Problem, die Rydberg-Ritz'sche Constante durch das Planck'sche h auszudrücken, hat mir schon lange vorgeschwebt... Wenn ich auch vorläufig noch etwas skeptisch bin gegenüber den Atommodellen überhaupt, so liegt in der Berechnung jener Constanten fraglos eine grosse Leistung vor." Sommerfeld to Bohr, 4 September 1913 (BCW 2: 603).

[11]Sommerfeld (1915). On Sommerfeld's work and its importance for the reception of Bohr's theory, see HDQT 1 (212–23), Heilbron (1967), Kragh (1985), Seth (2010). "Because of you, Bohr's idea has become completely convincing," Einstein to Sommerfeld, 8 February 1916; 3 August 1916 (DM).

of spectral lines due to an external electric field, a problem which Bohr considered crucial and was attempting to solve, too. Rubinowicz published calculations in 1918 that explained selection rules for the fine structure of hydrogen spectrum, which yielded results rivaling those achieved through Bohr's correspondence argument developed in Copenhagen, also in 1918 (Epstein 1916; Rubinowicz 1918). After the war, Sommerfeld could once again work with many advanced students from Germany and Austria, among them Walther Kossel, Adolf Kratzer, Karl Herzfeld, Gregor Wentzel, Pauli, and Heisenberg, whom he assigned various research problems in quantum theory (Eckert 1993).

Mindful of Germany's position in world science, Sommerfeld worried about postwar economic difficulties and especially about the scientific boycott organized by the victorious powers. He attempted to keep foreign contacts as active as possible by taking various opportunities to visit abroad, considering himself a diplomatic emissary of German science (Eckert 1993, Chap. 5). On his first such trip in September 1919, Sommerfeld went to Sweden, a neutral country that was the most sympathetic to Germany. The conference in Lund, organized by Manne Siegbahn, discussed research on X-rays, the field in which Sommerfeld was the foremost theoretical authority (Kaiserfeld 1993). Bohr, who was also in attendance, invited him to stop in Copenhagen on the way back, supported by the funds from the Danish Royal Academy.

The visit established a personal link and cooperation between the two professors and their research programs. Despite some remaining differences in approaches, both Bohr and Sommerfeld were pleased and in their subsequent letters addressed each other as friends. "[We] had long discussions about the general principles of the quantum theory, about which I think we agreed better when he left than when he came," reported Bohr to Rutherford. Besides scientific matters, the collaboration included mutual institutional support. It was decided that Sommerfeld's former assistant, Rubinowicz, who after the breakup of the Austro-Hungarian Empire lost his position at the University of Czernowitz, now in Romania, would come to Copenhagen. In November 1919, Sommerfeld informed Rubinowicz that Bohr would find him a one-year position. With the help of a grant from the Rask-Ørsted Fond, Rubinowicz worked in Copenhagen from April to August 1920 and again in May 1922 (Fig. 3.1).[12]

Upon returning to Munich, Sommerfeld received a letter from Harald Bohr with the request to help his brother's application for the purchase of a large spectrograph. The sum, 28,000 kr., was about ten times the usual amount of Carlsberg Fond's annual grants for professors, which apparently caused concerns. Sommerfeld's supportive letter to the foundation, official as it is, reflects his personal vision for Bohr's institute. In a time when the boycott did not allow Germany to maintain international research, Sommerfeld hoped to create a friendly offshore center. In Sommerfeld's view, the

---

[12]Bohr to Rutherford, 20 October 1919; Bohr to Ehrenfest, 22 October 1919; Rubinowicz to Bohr, 6 November 1919; Bohr to Sommerfeld, 19 November 1919. On his stay in Copenhagen, see also Rubinowicz to Epstein, 26 October 1919; 29 April 1920.

Fig. 3.1 Sommerfeld and Bohr at the 1919 meeting in Lund (NBA)

scientific interests of Germany and Denmark colluded, or at least complemented each other:

> The war burdens and unbearable peace terms have made scientific efforts in Germany impossible for a long time to come. Previously, Germany's numerous universities and institutes of technology were able to further experimental research with good financial support. Together with Germany, almost the whole European continent has become impoverished. But happy Denmark can step into the breach here. Denmark will enjoy doing this the more as such an act would honor the name of one of its outstanding sons. The institute of Mr. Bohr should not only serve the up and coming generation of Denmark, it will also be an international place of work for foreign talent whose own countries are no longer in a position to make available the golden freedom of scientific work. Just as in the past at the Vienna Institute of Radium Research, future researchers of all countries should meet one another in Copenhagen for special studies and to pursue common cultural ideals at the Bohr Institute of atomic physics.[13]

---

[13]Harald Bohr to Sommerfeld, 14 October 1919. "Durch die Kriegslasten und unerträgliche Friedensbedingungen ist es Deutschland, das bisher an seinen zahlreichen Universitäten und Hochschulen experimentelle Forschung mit reichen Mitteln gefördert hat, auf lange Zeit unmöglich gemacht, die Wissenschaft wie bisher zu pflegen. Zugleich mit Deutschland ist fast der ganze europäische Continent verarmt. Das glückliche Dänemark kann in die Bresche treten. Es wird dies um so lieber tun, als es dabei zugleich sich selbst in dem Namen eines seiner hervorragensten Söhne ehrt. Das Institut des Herrn Bohr sollte nicht nur dem dänischen wissenschaftlichen Nachwuchs dienen, es sollte eine internationale Arbeitsstätte auch für Talente des Auslands werden, denen die eigene Heimat nicht mehr die goldene Freiheit der wissenschaftlichen Arbeit gewären kann. Wie früher im Wiener Radium Institut so mögen künftig Forscher aller Länder zu besonderen Studien

For several years, until the break of their friendly relationship in 1923, Sommerfeld entertained the hope that the neutral Copenhagen institute would serve the interests of German science. The existing scientific disagreements with Bohr and the competition between the two research programs did not seem to him an obstacle. "I do not consider [Bohr] as foreigner,"—he once characteristically replied to Alfred Landé, who was trying to justify his hurry in publishing preliminary results of theoretical calculations by the need to secure German priority vis-à-vis Denmark.[14] Sommerfeld attempted to draw Bohr into the inner circle of his closest associates and former students with whom he was spending skiing vacations in the mountains. As early as the summer of 1918, when Sommerfeld was elected president of the German Physical Society, he suggested inviting Siegbahn or Bohr for a visit. This was hardly possible during the immediate postwar revolutionary chaos, but in March 1920 Max Planck sent an official invitation to Bohr to give a lecture to the German Physical Society in Berlin.[15]

3.3   New Partners

The general perception of overwhelming crisis—political, socioeconomic, and intellectual—was felt much sharper in Weimar Germany than in Great Britain, amplified by the bitterness over defeat in the war, the loss of prosperity, and a severe blow to academics' social prestige. In the period when German society as a whole was threatened by revolutions, military coups, and crises, science and its individual disciplines were also often declared to be in a state of deep crisis and ripe for radical conceptual changes. Even some older professors felt the need for new agendas, and even university administrators and ministry officials (at least in Prussia) wanted "to do new things" and to hire representatives of "the new physics."[16] By the latter they meant first and foremost research in the atomic and quantum domains, both theoretical and experimental.

---

in Kopenhagen sich treffen und im Bohrschen Institut für Atomphysik gemeinsame Culturideale verfolgen." Sommerfeld to the Carlsberg Fond, 26 October 1919, English translation in Robertson (1979, 34–35). On the Vienna Radium Institute and its dramatic decline after World War I because of economic hardships in Austria, see the correspondence between Stefan Meyer and Rutherford, in particular Meyer to Rutherford, 22 January 1920 (Eve 1939).

[14]"Ich sehe ihn nicht als Ausländer an," Sommerfeld to Lande, 31 March 1921. For the full text of the letter and analysis of its context, see Forman (1970, 214–17). For explanations of the differences and disagreements between Bohr's approach based on the correspondence argument and Sommerfeld's formal quantization rules, see Sommerfeld (1942), Darrigol (1992, Chap. 6).

[15]Sommerfeld to Bohr, 11 November 1920 (BCW 3: 690–91); 18 February 1921; Sommerfeld to Einstein, June 1918 (SWB); Planck to Bohr, 30 March 1920.

[16]For the pioneering and classic discussion of the cultural climate for science in Weimar Germany, see Forman (1971) and on the methodological and historiographic significance of Forman's approach for the history and sociology of science (Carson et al. 2011). On reformers in the Prussian Kultusministerium, see Forman (1967, 59–65), Born (1978, 200) and L. Lichtenstein to Sommerfeld, 14 January 1926 (DM).

The quantum concept originated in Germany during the first decade of the century and gained its initial international recognition shortly before the war. After the war's end, the topic suddenly became the most fashionable among German physicists and, within the field itself, the emphasis shifted from the earlier set of problems, dominated by thermodynamic methods, to new approaches: Einstein's light quanta, which had previously been largely ignored, and Bohr's atomic model. Radical shifts in research agendas occurred much faster than in postwar Britain, in part because of the accelerated and more pronounced changes not only in the academic community but in German society as a whole. In a wave of postwar professorial appointments, representatives of the younger generation were promoted to positions of influence. The social order in academia was shattered, just as it was in wider society, and the hierarchical structures inherited from the *Kaiserreich* era became less rigid. University reform made extraordinary professors and docents less dependent on ordinary professors, whose monopoly on financial resources was also weakened by the establishment in October 1920 of the *Notgemeinschaft der Deutschen Wissenschaft*, an independent source of research grants (Forman 1967: 59–100, 1974).

A symbolic event that revealed the softening of academic hierarchies occurred during Bohr's first visit to postwar Germany in April 1920. For the German Physical Society, it was one of the earliest meetings aimed at defying the international boycott and restoring contacts with other countries. Understandably, Bohr's lecture stirred much interest and the auditorium was crowded, with the crème de la crème of Berlin physics—Planck, Einstein, Haber, Nernst, Rubens, von Laue—sitting in the front row. Bohr, as was usual for him at formal presentations, spoke badly and very quietly; it was difficult to hear him and almost impossible to understand his German. After the talk, a group of younger scientists who sat further back and could not hear much approached the speaker and asked him to repeat the presentation the following day to a different audience, with no ordinary professors (Levi 1985, 47). Such a display of independence demonstrated by this so-called bosses-free congress (*Bonzenfreier Kongress*) could have happened in the earlier founding years of the German Physical Society around 1845, but not during the height of imperial power between 1870 and 1918. The colloquium photo shows Bohr surrounded by the up and coming generation of German physicists, for many of whom he became an inspiration and unofficial role model. "Bosses-free" was to some degree an overstatement, since participants—eighteen total—were already accomplished researchers in their own right. Their career advancement, unlike Bohr's, had been delayed by the war, but they were already on the verge of receiving professorships. One of them, James Franck, would indeed become a professor at the University of Göttingen less than one year later and, together with another new professorial appointee, Max Born, started developing a new major center for quantum physics (Fig. 3.2).

Together with Hertz, Franck had opened up a new experimental field, the study of quantum collisions between atoms and electrons, with their famous experiment of 1913 (Hon 1989; Gearhart 2014). They discovered discontinuous jumps in the amounts of energy transferred from electrons to atoms, at first interpreting the results as a measure of ionization potential. Although aware of the Bohr model, they did not initially use it to interpret their experimental findings. After 1917, upon resumption

Fig. 3.2 "Bosses-free congress" in Berlin–Dahlem, April 1920. Otto Stern, Wilhelm Lenz, Franck, Rudolf Ladenburg, Paul Knipping, Bohr, Ernst Wagner, Otto v. Beyer, Otto Hahn, Lise Meitner, Hevesy, Wilhelm Westphal, Hans Geiger, Gustav Hertz, and Peter Pringsheim (NBA)

of his research interrupted by war service, Franck's views changed. He now understood energy jumps to be a measure of excitation potentials and, as such, a direct confirmation of the Bohr theory of discontinuous energy levels in atoms (Franck and Hertz 1919; Lemmerich 2011). In 1919–1920, Franck worked at the *Kaiser Wilhelm-Institut für Physikalische Chemie und Elektrochemie* in Berlin on the exact determination of energy levels in helium. These measurements were important for Bohr and Kramers's attempts to extend the quantum model of hydrogen to helium, the next simplest atom in the periodic table, but with a much more complex spectrum.

Bohr had Franck in mind as he searched for an experimental physicist for his new institute but learned that Franck had just accepted a Göttingen professorship. Funds were available from a private source (probably the Berlème committee) to invite a distinguished foreign physicist to the official opening of the Copenhagen institute in March 1921. Einstein was sought but had to make his apologies. Franck gladly accepted Bohr's invitation to visit in early 1921, prior to taking up his duties in Göttingen, and to help Bohr jump-start experimental research. He spent February and March in Copenhagen, attended the opening ceremony, gave a lecture at the Danish Natural Science Society, and advised Jacobsen, a student of Hansen, on how to set up an experimental apparatus for the study of atomic collisions. Although

Frank's stay was too short for conducting actual experiments, he expressed interest in visiting again in 1922.[17]

The story of multiple professorial appointments in physics at the University of Göttingen in 1914–1921 was so bureaucratically complicated that even seasoned officials at the Prussian educational ministry had problems straightening it out. Toward the end of a long process, in April 1920 Max Born was offered the ordinary professorship in experimental physics. A mathematical physicist par excellence, Born hesitated to accept the duties of teaching and directing experimental research. While bargaining with the ministry over the terms of the appointment, he exploited a confusion in the official paperwork to suggest the possibility of hiring a second extraordinary professor in addition to the existing one, Robert Wichard Pohl. At first, the bureaucrat

> laughed, saying that this was obviously an error of the copyist; but he soon saw the possibilities… [W]e were still living in revolutionary times. Like most of the officials in the Ministry of Education, he was a new man, keen to do new things (Born 1978, 200).

Born's wish to accept the duties of professor of experimental physics only nominally, and to delegate them in practice to two subordinate extraordinary professors, eventually resulted in the establishment of three full professorships in place of one. Both Pohl and Franck were promoted to professors of experimental physics and directors of separate university institutes, while Born's title renamed to the ordinary professor of theoretical physics.[18]

Freed by this arrangement from obligations toward experimental research, Born could concentrate on theory. He had applied quantum concepts to crystal lattices even prior to 1914. Restarting research in 1918 after military service, Born shifted to quantum models of atoms and molecules, which moved him closer to the field occupied by Bohr and Sommerfeld. In his new Göttingen location, he aspired to establish another "school of theoretical physics" in addition to the one that already existed in Munich under Sommerfeld. Until he could train his own doctoral students, Born had to borrow assistants from elsewhere. In 1921, his first year in the office, Born relied on his predecessor's student Erich Hückel, a Hungarian visitor Ernst Brody, and Sommerfeld's recent PhD Wolfgang Pauli. As Sommerfeld was leaving for a lecture tour in the USA in September 1922, he agreed that Heisenberg, still a doctoral student in Munich, would come to Göttingen for an academic year. Later, Born's own students obtained doctorates with him, including Friedrich Hund (1922), Lothar Nordheim (1923), and Pascual Jordan (1924). By 1928, writing in a *Festschrift* for Sommerfeld and praising the latter as the "founder of a school," Born could already feel himself on a similar footing, offering a gesture of appreciation from the head of another important and rival school of theoretical physics. In May 1922, Born

---

[17]Bohr to Franck, 18 October 1920; 29 January 1921; Franck to Bohr, 15 April 1921.

[18]Somewhat different descriptions of bureaucratic struggles to choose a successor to Debye in Göttingen are presented in IMN 2 (301–02, 356–57), HDQT 1 (292–94), Hund (1987), Dahms (2002). On Born's appointment as *Ordinarius* for experimental physics, and eventually as professor of theoretical physics, and on his quantum research program, see Greenspan (2005), Schirrmacher (2019).

was still at the very beginning of the process when he wrote to Sommerfeld about his plans to allow his people "to quantize to give you a little competition."[19]

Bohr's contacts with Born started in June 1922 during a series of seven lectures on atomic theory he delivered in Göttingen, the ten days that became known as the "Bohr Festival."[20] When Bohr received his Nobel Prize several months later, "he became a kind of idol," so Hund described Bohr's status among Göttingen physicists (Hund 1985, 73). Franck recognized Bohr as the leader and, along with Einstein, the main authority in physics even before 1922. In his Göttingen location, he was eager to work experimentally on Bohr's agendas and valued his contacts with Copenhagen as "fertilizing for a field."[21] Franck frequently asked Bohr for scientific advice and sent him manuscripts for approval prior to submitting them to print (a behavior typical of a student, but not of an independent professor). He also arranged for Paul Hertz's translation of Bohr's seminal early papers into German (Bohr 1922), paid by the Rask-Ørsted Fond as part of the program to promote Danish research to foreign audiences.[22] At first, Franck's dependence on Bohr annoyed Born: "we had discussed a problem thoroughly and come to a conclusion. When I asked him after a while: 'Have you started to do that experiment?' he would reply: 'Well, no; I have written to Bohr and he has not answered yet.' This was at times rather discouraging for me, and even retarded our work to some degree" (Born 1978, 211). But when Born wrote to Copenhagen for the first time, six months after Bohr's visit to Göttingen, he did not mention competition, like he did with Sommerfeld, but politely asked Bohr's permission to enter the field of quantum atoms. Born enclosed with his letter the manuscript of a paper he had written jointly with Heisenberg and inquired "whether you think it right to publish the result," adding deferentially: "I would not like to publish anything on this, your very own areas of work, which could not meet with your approval."[23] Unlike Sommerfeld, whose attitude toward Bohr was relatively patronizing, and despite their nominally high status in the German physics community, until 1926 the Göttingen quantum physicists positioned themselves as Bohr's, but not Sommerfeld's, junior partners.

---

[19]Born (1928), HDQT 1 (361–63), Born to Sommerfeld, 13 May 1922, quoted in IMN 2 (357–58).

[20]On behalf of the Göttingen physico-mathematical seminar, David Hilbert invited Bohr to be their first postwar visiting professor in the summer term of 1921. Because of an illness and overwork, Bohr postponed his planned visit until the following year. Hilbert to Bohr, 10 November 1920; 11 November 1920; 18 April 1921; Bohr to Hilbert, 22 November 1920.

[21]Franck to Bohr, 29 July 1922; 23 December 1922.

[22]Bohr to Franck, 16 September 1921; 27 September 1921; Hertz to Bohr, 29 September 1921; Bohr to Hertz, 8 October 1921; 14 November 1921; 26 January 1922.

[23]Franck to Bohr, 25 September 1921; 29 September 1921; 21 February 1922. Born to Bohr, 4 March 1923: "ob Sie es für richtig halten, das Resultat zu publizieren... Ich möchte auf diesem, Ihren eigenen Arbeitsgebiete, nichts publizieren, das von Ihnen nicht gebilligt werden könnte."

3.4 The Value of Neutrality

On 31 January 1920 Bohr sent his first application to the Rask-Ørsted Fond requesting support for three foreign visitors: Hevesy from Hungary, Rubinowicz from Poland, and Klein from Sweden.[24] Similar geographical patterns characterized the early years of the institute, until 1924. Bohr arranged fellowships for physicists who typically came from either neutral countries (besides Kramers, Klein, and Rosseland, also Dirk Coster from Holland, and Werner Kuhn from Switzerland) or from countries that formerly belonged to the Austro-Hungarian Empire (besides Hevesy and Rubinowicz, also Pauli, who was Austrian and was referred to as such when his visit to Copenhagen was arranged). In the former Habsburg lands, the postwar economic situation and conditions for scientific work had become so miserable—much worse than anywhere else in Europe, with the exception of civil war-torn Russia—that even modest conditions in Copenhagen seemed prosperous by comparison.

Besides the above two groups, visitors from the USA and Japan arrived in Copenhagen with some regularity, as elsewhere in Europe, but typically with their own funds. Bohr did not have to secure financial support for them and, apparently, did not concern himself too much with overseas students during the initial years of his institute. As for representatives from major European countries such as Britain, Germany, or France, they were not yet coming to Copenhagen for either extended work or study. A few distinguished colleagues appeared on special invitations for short visits: Rutherford and Owen Richardson from Britain, Ehrenfest from the Netherlands, Sommerfeld, Landé, Franck, and Paschen from Germany. Bohr had the means to invite, on average, one such honorary guest per year to give a lecture and to discuss scientific and institutional topics. Younger students and assistants from Germany would not start working in Copenhagen until 1924, i.e., after hyperinflation that undermined their financial conditions at home and after the start of the Rockefeller program of international postdoctoral fellowships.

The presence of neutrals and former Austro-Hungarians in the Bohr institute established its early international profile and helped create a network of scientific contacts, both of which played an important role in the 1922 discovery of hafnium by Coster and Hevesy. Coster learned about opportunities to do research in Copenhagen from his fellow countryman Kramers. In 1920, he was planning to go abroad for experimental studies of X-rays. After the British refused to grant him a visa, Coster wrote to Bohr asking whether he could come to Denmark if neither Debye in Zürich nor Siegbahn in Lund accepted him. He did in fact go to Lund, but Bohr invited him to visit Copenhagen afterward.[25] In summer 1922, Coster defended his dissertation in Leiden on "Röntgen Spectra and Bohr's Atomic Theory." Meanwhile, Bohr also developed an interest in X-ray spectroscopy and, in anticipation of Coster's visit, raised funds from the Rask-Ørsted Fond and purchased an experimental apparatus.[26]

---

[24] Bohr to Rask-Ørsted Fond, 31 January 1920 (NBA).

[25] Radder (1982, 226); Coster to Bohr, 11 July 1920; Bohr to Coster, 20 July 1920.

[26] Bohr to Coster, 3 July 1922; Coster to Bohr, 3 August 1922.

Bohr turned to X-ray spectra in connection with his newest theory of the periodic table of chemical elements. His and Kramers's persistent attempts to calculate quantum orbits for atoms beyond hydrogen had not achieved much success since 1919. In 1921, Bohr changed the approach and, instead of exact models for specific atoms, suggested in two letters to *Nature* a general qualitative explanation for the entire periodic table, lengths of its periods, and patterns of electronic configurations in different groups (Bohr 1921). The theory was based on intuitive reasoning that combined various spectroscopic and chemical data, but Bohr conveyed the impression that it could also be derived more strictly from his correspondence principle. German quantum theorists were excited and anxiously awaited Bohr's June 1922 lectures in Göttingen, hoping to learn how one could derive the new theory quantitatively from basic quantum postulates. Unfortunately, Bohr was not in a position to offer precise mathematical calculations, and even his qualitative theory met a serious challenge. Upon returning from Göttingen, he learned from Rutherford's note in *Nature* that Alexandre Dauvillier and Georges Urbain in France claimed to have discovered one of the few missing chemical elements of the atomic number 72, which they called celtium and classified as belonging to the group of rare earths. In Bohr's scheme of the periodic table, the rare earth group was supposed to end with the element 71. Bohr thus had to acknowledge the apparent contradiction between his theory and the existing experiments in a letter to Franck and in an appendix to the 1922 German translation of his essays.[27]

Coster was preparing for his Copenhagen visit and upcoming work on X-ray spectra of rare earths when he received a letter from Bohr inquiring about the reliability of the French results. Coster expressed serious skepticism, which only increased after Siegbahn visited Paris and was shown photographic plates of dubious quality. Coster even composed a critical note to *Nature*, but Bohr did not find it advisable to rush to publication and wrote a personal letter to Rutherford instead, explaining his persisting doubts in the discovery of celtium.[28] Coster arrived in Copenhagen in September 1922 and started, together with Hevesy, an experimental investigation that by the year's end was able to report a major discovery. They found the new element 72, called hafnium (after the Latin name of Copenhagen), in zirconium ores and, in agreement with Bohr's predictions, it was not a rare earth, but a chemical analog of zirconium. The news arrived just in time for Bohr to include it, at the very last minute, into his talk at the Nobel ceremony on 11 December, when he was awarded the 1922 Nobel Prize in physics "for his services in the investigation of the structure of atoms, and of the radiation emanating from them" (Coster and Hevesy 1923; Kragh 1979, 178–86).

In the heated atmosphere of postwar tensions, the priority conflict could not avoid being understood and interpreted from nationalistic positions. The French side of the dispute was represented by the chemist Urbain, the physicist Dauvillier, an expert on

---

[27]Bohr (1922). On Bohr's theory of the periodic system and chemists' and physicists' diverging views on the element 72, see Kragh (1979, 2012, Chap. 7), Scerri (1994).

[28]Bohr to Coster, 3 July 1922; 5 August 1922; Coster to Bohr, 15 July 1922; 16 July 1922; 28 July 1922 (BCW 4: 674–8).

X-rays Maurice de Broglie, and French journals, which refused to discuss the Danish claim. The actual debate took place in British scientific magazines, whereby some British chemists and journal editors supported their French allies, while Rutherford took the side of Bohr and the neutrals. The Copenhagen discovery, although formally made in a neutral country, was still suspected to represent a pro-German case. After all, Hevesy came from Austria, while Coster held strongly pro-German views. Moreover, the Copenhagen team received informal assistance from the Austrian chemist Carl Auer von Welschbach, who supplied Copenhagen with his chemical samples of rare earths and who had had a long, prewar record of poisonous clashes with Urbain over chemical elements 70 (ytterbium/cassiopeium) and 71 (lutecium/aldebaranium). Privately, Coster expressed the hope that the hafnium debate would also reopen Auer's earlier priority claims that had been rejected by the "international" (Coster's dismissive quotation marks) committee on atomic weights, headed, as a matter of fact, by Urbain. The actual proposal to look for element 72 in zirconium ores came from Hevesy's Austrian colleague and friend Fritz Paneth, then working in Berlin, who also helped to mobilize German scientific periodicals in support of the Copenhagen discovery and of the name hafnium.[29]

On Coster's suggestion, Bohr contacted Auer von Welschbach, requested chemical preparations, and sent in return a financial contribution for the physical institute at the *Technische Hochschule* in Vienna. As in other delicate matters, Bohr was very careful in words and avoided any open mention of the related nationalistic sentiments.[30] This does not mean that he did not think of politically controversial questions; just the opposite is probably true—that he gave them much consideration and therefore was particularly cautious in his own statements. Bohr's position can be sensed from his scarce remarks and subtle actions in response to his correspondents, whose letters, actions, and motivations were usually much more open. In his early postwar correspondence with British colleagues, Bohr tried to draw attention to the dire economic situation in Germany and to entice some sympathy toward the defeated side. He felt "quite sure that the men now in power in Germany take a real peaceful attitude, not for the occasion, and not because they have always done so, but because all < ...minded > people in the world seem to have understood the unsoundness of the principles on which international politics has hitherto been carried on." In his opinion, German political prospects "seem to take a happier turn" after the lost war and, "if... only there will be no anarchy in Germany due to the great need and poverty at the present moment, their future may certainly be looked upon as the beginning of a new era in history."[31]

The Danish academic community had its own disagreements with regard to the international tensions in postwar science. Knudsen, in particular, supported the new international organizations that excluded Germany and Austria from membership.

---

[29] Kragh (1980), Robertson (1979, 69–73); Coster to Auer, 24 January 1923; 14 March 1923; 27 March 1923.

[30] Bohr to Auer, 5 July 1922; 25 September 1922; 11 December 1922; Auer to Bohr, 12 September 1922.

[31] Bohr to Richardson, 25 January 1919; Bohr to Rutherford, 24 November 1918.

By comparison, Bohr appears to have preferred more symmetrical relationships with the war's winners and losers. In 1924, he received an invitation to participate in the Solvay conference in Brussels from Hendrik A. Lorentz, who wrote apologetically regarding the continuing boycott of German scientists:

> After all that had taken place in Belgium, Germans could not be invited to Brussels three years ago, nor can they be now. We must have patience and trust in the future. But I am convinced that the ultimate aim will best be reached if those who wish to attain... do not stand aloof, but, taking their share in what can be done, try to work in a spirit of reconciliation and so to pave the way for a better understanding between the nations that have now been estranged from one another.

Bohr did not come personally to the 1924 Solvay meeting, nor, probably for political symmetry, to the special panel on atomic science at the meeting of the *Gesellschaft Deutscher Naturforscher und Ärzte* in Innsbruck, Austria (Kramers delivered the reports on their research in Copenhagen). In both cases, Bohr excused himself politely with references to the extra work on the planned enlargement of his institute and various other occupations.[32] He accepted an invitation to attend the next Solvay congress in 1927, and the first in which the Germans were allowed to participate. It is not accidental that in the period from 1918 to 1926 his institute published almost equal numbers of articles in English (93) and in German (95), with a much smaller number of papers published in Scandinavian languages.[33]

In retrospect, Bohr's position toward the contested models of postwar internationalism appears to be one of the carefully balanced neutrality and mediation between the two hostile academic camps. In the specific situation of international boycotts, this meant, in practice, a relatively pro-German course and was often perceived as such. What was probably even more important in those political and economic circumstances was that physicists from Germany and former Austro-Hungary were much more interested in contacts with neutrals than their British or French colleagues. The potential of possible collaboration was also huge, which Bohr noticed firsthand during his visit to Germany in April 1920 and subsequently reported in a letter to a British colleague: "It was a very interesting event for me having not had the opportunity of meeting Planck and Einstein before... although it was a sad experience to witness all the poverty and depression, which, however, did not seem to interfere much with the scientific activity in Germany, which for the moment is as intense as ever if not more."[34]

---

[32]Lorentz to Bohr, 16 February 1924; 4 June 1926; Bohr to Lorentz, 3 March 1924; 2 July 1925; 24 June 1926; Rassow to Bohr, 29 January 1924; Bohr to Rassow, 28 March 1924 (BGC).

[33]In the period from 1927 to 1934, the balance would shift more toward the German language: 74 versus 40 publications, and from 1935 to 1941, to the English: 153 versus 24 publications. Bound collection of institute's reprints *Afhandlinger*, Universitetets Institut for Teoretisk Fysik (NBA).

[34]Bohr to Richardson, 12 June 1922.

Chapter 4
A "Kuhnian" Crisis

Back when Thomas Kuhn's model of scientific revolutions was popular among historians of science, they applied it to numerous case studies to test, which aspects of the model could be confirmed or refuted by historical evidence. The results were typically ambiguous: Due to flexible criteria and the complexities of any real story, historians could identify as many supporting examples and as many counterexamples as they wished. Yet one puzzling aspect of those efforts stood apart. While it was not difficult to find occasions on which scientists congratulated themselves on having achieving a "revolution" in some area of research, much less readily did those scientists openly admit that their own field was in a state of "crisis" and conceptual chaos. The prolonged crisis stage in Kuhn's model—when the common paradigm is lost and the professional community is searching, in confusion, for possible future alternatives—could be identified, imagined, or constructed post-factum by later generations of scientists or historians, but very rarely by participants themselves. Among a few exceptions, as far as I know, none is closer to Kuhn's description than the so-called crisis of the old quantum theory in 1923–25, when some of the major authors and leaders of that theory openly questioned its conceptual foundations.[1]

We can now reverse the question. Kuhn, before he turned into a philosopher and historian of science, had written a PhD thesis in quantum spectroscopy under John Van Vleck and certainly knew the field's historical mythology as told and retold by its senior practitioners. It thus looks possible, and even plausible, that the Kuhnian model of scientific revolutions came about as a generalization of a particular disciplinary memory, the story of a major crisis in quantum theory prior to the great revelation of quantum mechanics. This chapter inquires how, and among whom, that state of dissatisfaction and disorientation developed by 1925.

---

[1] Historians have also identified the limits of that discourse: It was localized mostly in Göttingen and Copenhagen, among scientists connected with these two centers, but not so much in Munich, and even less so in remote peripheral locations (Seth 2010, Chap. 7).

A. Kojevnikov, *The Copenhagen Network*, SpringerBriefs in History of Science and Technology, https://doi.org/10.1007/978-3-030-59188-5_4

4.1 The Field in Disarray

Not only local physicists and mathematicians attended Bohr's Göttingen lectures in June 1922. Sommerfeld arrived from Munich, bringing along his promising new student Heisenberg, Ehrenfest travelled from Leiden, Landé from Frankfurt, and Pauli from Hamburg. All were anxious to learn about Bohr's calculations behind his quasi-intuitive theory of the periodic system of elements and also his opinions on the growing number of difficulties and disagreements within the quantum theory of atoms. In the course of one of Bohr's lectures, Heisenberg stood up and voiced objections, not merely because he was so young and arrogant, but because he expressed the attitude and had the backing of his supervisor, Sommerfeld.[2]

This was Heisenberg's first personal encounter with Bohr. In response, Bohr reportedly took him out for a long walk in the woods, and their probing philosophical dialogue, according to the popular historical myth that Heisenberg himself helped proliferate later, left a deep and profound influence on the young student. Letters and papers from that period, however, do not confirm this interpretation. For another year and a half, until 1924, Heisenberg would continue to be Sommerfeld's loyal disciple, doing what his teacher expected from him, working on the latter's agenda, and expressing critical opinions of Bohr's. On one of those days in Göttingen, Bohr, Sommerfeld and Heisenberg privately discussed the subject of their most important disagreements regarding multiplets in the spectra of atoms with more than one electron and the complex splitting of lines in the magnetic field, or what was then called the "anomalous Zeeman effect." Experimental data came mainly from the spectroscopic laboratory of Paschen and Ernst Back in Tübingen. In 1921, Landé in Frankfurt proposed a successful phenomenological classification of the multiplets, known in later textbooks as the vector model of multielectron atoms, which required him to introduce quantum numbers with heretofore unknown half-integral values (Forman 1970). Although disagreeing with Landé on some aspects, Sommerfeld was eager to incorporate his achievement into the basic quantum theory of the atom. He assigned the work to a new student of his, and in late 1921 Heisenberg developed what became known as the "core model" of the multielectron atom that explained Landé's classification at the price of abandoning some basic postulates of the Bohr-Sommerfeld atomic theory (Cassidy 1979). To achieve agreement with known experimental data, Heisenberg had to assume the existence of magnetic moments inside the atom core and of half-integral quantum numbers. Sommerfeld was somewhat concerned about the proposed bold deviations from his quantization rules, but valued the overall accomplishment and wrote with excitement about his new pupil:

> I expect tremendous things from Heisenberg, probably the most gifted of all my students, including Debye and Pauli. His Zeeman model generally meets with opposition, particularly

---

[2]Heisenberg later recalled "We could clearly sense that he had reached his results not so much by calculation and demonstration as by intuition and inspiration, and that he found it difficult to justify his findings before the Göttingen's famous school of mathematics." (Heisenberg 1971, Chap. 3). For the content and detailed discussion of Bohr's lectures and explanation of Heisenberg's objections, see (BCW 4: 341–419; HDQT 1: 332–58; 2: 128–29).

from Bohr. But I find its success so enormous that I held back all my reservations about its publication... Heisenberg is only in his fourth semester and 20 years old.[3]

Having heard a rumor that Bohr preferred a different explanation of the multiplets, through electrical rather than magnetic interaction, Sommerfeld sent him the proofs of the forthcoming paper by Heisenberg: "I certainly do not wish to disturb your circles or delay your final publication! But it seemed right to me that you set your eyes on our assertions as early as possible." Bohr did not have a developed theory of his own, but he strongly objected to Heisenberg's: "My point of view is that the entire method of quantization (half-integral quantum numbers and the rest) does not appear reconcilable with the basic principles of the quantum theory, especially not in the form in which these principles are used in my work on atomic structure."[4] In his own work, Bohr agreed to sacrifice many fundamental principles of classical theories, but was reluctant to tinker with the quantization principle, which he himself had introduced into physics in 1913. On top of that, Heisenberg's model did not reconcile with Bohr's "building-up" approach, which he used in 1921 to interpret the periodic system of elements (Fig. 4.1).

In addition to disagreements regarding the anomalous Zeeman effect, another difficulty emerged from calculations of the helium atom. Kramers, under Bohr, had been attacking the problem since 1919, but despite enormous mathematical efforts spent to design a model with orbits for two electrons, had not achieved a clear success (Pais 1991, 197). Bohr mentioned Kramers's calculations in several presentations but was not ready to publish, judging the result "not very satisfactory." In Göttingen, he learned that Pauli, working with Born in early 1922, also calculated the helium model and its spectra, obtaining numerical results that were in apparent disagreement with experimental data. Bohr communicated the negative result to Kramers and urged him to complete a paper, since "we cannot wait much longer with publishing our calculations of the helium orbits."[5] Bohr offered Pauli an assistant position for one year in Copenhagen, the duties of which included helping with calculations and with writing Bohr's papers in German. On his return journey, he stopped in Hamburg to meet Wilhelm Lenz at the railway station. Pauli had been working as Lenz's assistant since April 1922 and needed his professor's permission to come to Copenhagen in September.

Pauli assured Bohr that he did "not intend at all to get in [Kramers's] way with a publication" and, indeed, he did not publish his calculations of the helium spectra.

---

[3]"Von Heisenberg, der wohl der begabteste all' meiner Schüler ist, einschl. Debye und Pauli, erwarte ich Ungeheueres. Sein Zeeman-Modell stösst auf allgemeinen Widerspruch, besonders auch bei Bohr. Ich finde aber den Erfolg so gewaltig, dass ich alle Bedenken bei der Publikation zurücksteckte... Heis[enberg] ist erst im 4n Semester und 20 Jahre alt." Sommerfeld to Epstein, 29 July 1922.

[4]"Ich möchte gewiss nicht Ihre Kreise stören oder Ihre endgültige Publikation verlangsamen! Aber es schien mir richtig, dass Sie unsere Behauptungen so früh wie möglich zu Gesicht bekommen." Sommerfeld to Bohr, 25 March 1922; Bohr to Landé, 15 May 1922. On discussions in Göttingen and Bohr's opposition to Heisenberg's magnetic core model, see Cassidy (1979, 216–17, 1992, Chap. 7).

[5]HDQT (1: 408–9); Bohr to Landé, 26 June 1919; Bohr to Kramers, 15 July 1922.

Fig. 4.1 Sommerfeld, Heisenberg, Oseen, Ladenburg, Stern, Hertha Sponer, Günther Cario in Göttingen, 16 June 1922

While Kramers was completing his paper on helium, which was finished by the year's end (Kramers 1923), Bohr put Pauli to work on the problem of the anomalous Zeeman effect, a novel topic for Copenhagen. Pauli's task was to design an alternative to Heisenberg's theory and defend strictly integral quantum numbers against the heresy of the core model. In early 1923, three manuscripts were ready: Pauli's technical response to Heisenberg, which he sent to the latter for comments; Bohr and Pauli's general discussion and interpretation of Landé's results, sent to Landé; and a note by Kramers and Pauli on diatomic molecules with an alternative to an earlier paper by Adolf Kratzer, another pupil of Sommerfeld. "As you will see from our note, we are always so conservative here in Copenhagen that we stick sincerely to the integral quantum numbers," Bohr confessed to Landé the main goal of these efforts. Landé, in fact, liked neither of the two sophisticated mathematical attempts to explain his phenomenological vector model. He found Sommerfeld and Heisenberg's theory "certainly wrong" (*sicher falsch*), while Bohr and Pauli's "artificial, simply to avoid half-quanta" (*künstlich, um halbquanten zu vermeiden.*)[6]

Though certainly a mathematical virtuoso thoroughly trained in Sommerfeld's formal methods, Pauli proved far less successful in defending integer quantum numbers than Coster and Hevesy in defending Bohr's views on the periodic system. Heisenberg had not expected such strong opposition from Bohr's camp to his core model, but he was not easily discouraged. Upon receiving Pauli's manuscript and letters, Heisenberg criticized their results strongly in a letter to Sommerfeld, who was away on a trip to America, adding a pun that Pauli in Copenhagen had become "too bohred" (*sehr "verbohrt"*). Heisenberg's critique and Landé's additional progress in classification of the spectra with the help of half-integral numbers convinced Bohr and Pauli to withdraw their respective papers from publication. Of the three Copenhagen manuscripts, only the paper by Kramers and Pauli (1923) appeared in print. In one letter to Landé, Bohr reluctantly admitted defeat by accepting a restricted possibility of half-quanta, but in print he preferred to explain away the difficulties by hinting at some unspecified mysterious "non-mechanical force" (*unmechanischer Zwang*), which, in Daniel Serwer's appropriate characterization, "was nothing more than the cause of inexplicable occurrence."[7]

Meanwhile, in Göttingen Born and Heisenberg applied advanced methods of perturbation theory to achieve a more general and thorough treatment of the helium problem and a more definitive negative verdict: "one cannot arrive at an explanation of the helium spectrum by consistent application of the known quantum rules." To explain the failure, Born presented a dilemma: either in the helium case, too, the rule of quantization with strictly integral quantum values had to be modified (the proposal favored by Sommerfeld and Heisenberg), or the laws of mechanics no longer held for describing even the stationary states of the atom. Both options violated either one or the other fundamental postulate of Bohr's original atomic theory. Born reported "the catastrophe" in a letter to Bohr, enclosing the manuscript to inquire about Bohr's approval. In Copenhagen, Pauli and Kramers checked and confirmed the correctness of the calculations. Bohr judged the negative result "very important" and looked for an explanation in "the inapplicability of the current foundation of quantum theory insofar as it concerns systems with multiple electrons." In the case of helium, as in the case of the anomalous Zeeman effect, he could offer only a verbal rationalization of the failure by saying that quantum theory in its existing form was not applicable to atoms with more than one electron, that is, to all atoms except hydrogen and ionized helium.[8]

---

[6] Pauli to Bohr, 7 July 1922; 5 September 1922; Bohr to Landé, 14 February 1923, English translation from HDQT (1: 486); Landé to Epstein, 1923.

[7] Heisenberg to Sommerfeld, 4 January 1923; 14 January 1923, cited in HDQT (1: 491); Bohr to Landé, 3 March 1923; (Serwer 1977, 191; Cassidy 1979, 220–222).

[8] Born and Heisenberg (1923, 229), English translation from Duncan and Janssen (2019, 382); Born to Bohr, 4 March 1923; 11 April 1923; "Unzulässigkeit der jetzigen Grundlage der Quantentheorie, soweit es Systeme mit mehreren Elektronen betrifft." Bohr to Born, 2 May 1923; (Darrigol 1992, 175–9; Serwer 1977), Pauli to Landé, 23 May 1923 (PWB, 90).

Pauli was depressed: "How can one look happy when he is thinking of the anomalous Zeeman effect?"—he recalled having said to a colleague at the time. "Pauli thought that the [Heisenberg] core model was fundamentally incorrect, but he could offer no alternative. Measured against his own standards, Pauli's work in Copenhagen seemed to be a failure," concludes Serwer. Pressed by Bohr, in April 1923 Pauli submitted his rather meager results for publication, in which he tried to avoid constructing any specific atomic models when discussing the spectral multiplets. After Pauli left Copenhagen in October 1923, he summarized his disillusionment in a letter to Bohr in the following words:

> The atom physicists in Germany [he could have added Denmark - AK] fall today into two groups. One group first works through a given problem with half integral values of the quantum numbers and if it does not agree with experience then they work it with integral quantum numbers. The others calculate first with integral numbers and if it does not agree, then they calculate with halves... I myself have no taste for this kind of theoretical physics and retire from it to my heat conduction in solid bodies.[9]

Pauli's negative experience with the anomalous Zeeman effect and helium led him to question the usefulness of atomic models in general, not just any particular kind of them, and made him increasingly skeptical of applying mechanical pictures, orbital trajectories, and classical analogies to the description of the atom interior (Hendry 1984, 49–50). Approximately one year later, he returned to the problem of complex spectra and developed a different proposal based on Landé's final improvement of the vector model and Edmund Stoner's 1924 explanation of the periodic system of elements, rather than the one by Bohr. His new theory included half-integral values but reassigned them as an additional, fourth quantum number for the electron, rather than the atomic core (*Zweideutigkeit*, now famous as Pauli's exclusion principle) and stood in a somewhat ambiguous relationship to the Copenhagen legacy: Bohr's "building-up principle" was preserved, but not the correspondence principle. Instead of emphasizing analogies with classical models, Pauli started professing "a radical sharpening of the opposition between classical and quantum theory."[10]

Upon his return from America in May 1923, Sommerfeld reported to Epstein that "Bohr has capitulated with regard to the Zeeman effect." By that time, the once friendly relations between Munich and Copenhagen had strained. Some scientific disagreements had been present all along—diverging views on Bohr's correspondence principle, Sommerfeld's inner quantum number *j*, and the wave–particle issue in radiation theory—but were tolerated for the sake of collaboration.[11] By early 1923, however, Bohr was already voicing his criticism of Sommerfeld's proposals freely, while Sommerfeld was openly rejoicing over various failures of Bohr's attempts,

---

[9]Pauli to Bohr, 21 February 1924 (PWB, 147–8, English translation in Serwer 1977, 228).

[10]Pauli (1925), Darrigol (1992, 201–9), Enz (2002, 106–9). "radikalen Verschärfung des Gegenzatzes zwischen klassischer und Quanten-Theorie," Pauli to Sommerfeld, 6 December 1924 (PWB 182–83). For Pauli's excuse for not using the correspondence principle, see Pauli to Bohr, 12 December 1924 (PWB, 188).

[11]"Bohr hat im Bezug auf die Zeeman Effekte kapituliert." Sommerfeld to Epstein, 5 May 1923; (Darrigol 1992, Chap. 6).

hinting at not only scientific, but also personal and political problems—his bitterness over the fact that the 1922 Nobel Prize was awarded to Bohr alone rather than shared by the two of them, and his disappointment in the hopes that Bohr's institute would work for the interests of German science:

> I have not congratulated you for the Nobel Prize yet; I do so now, warmly and sincerely, on the several-days-long journey from Chicago to Pasadena... Hopefully the overall and early recognition of your work will help you... overcome occasional disagreements... I do not want to tell you about America, only that I have been received cordially everywhere (*almost* everywhere). Neither do I want to lament to you about the German misery, although I, observing from afar with my hands tied, feel it twice as strongly. I would rather speak to you of a few scientific experiences. On my very first day in Madison I spoke with young Van Vleck... about his calculation of the Bohr-Kemble He model. Now there can be no doubt that this model and its *horrible* conception of magnetism is wrong... Here I am lecturing everywhere about your new atomic model, which I explain with the help of color slides. Of course, I replace your K shell with my view of the He model.

Science, politics, institutional and personal rivalry all intermingled in these disagreements. As he was writing this letter, Sommerfeld did not yet anticipate that his former student Pauli, then on a visit to Copenhagen, would soon refuse to return to Munich for *Habilitation*. But Sommerfeld certainly understood that hyperinflation, which was skyrocketing in Germany during the fall of 1922, was weakening the position of German science even further. He ended his letter with bitter words that practically amounted to the discontinuation of correspondence and of the relationship:

> If you want to write me, it would be better to address it to... But you *do not have* to write to me. Enjoy your young fame in a well-ordered Fatherland, and in the bosom of a loving family! And think about how miserable things are for us, Germans![12]

Bohr was thinking about the situation in Germany, but he did not answer Sommerfeld's letter. By that time, Göttingen had already replaced Munich as his primary German partner. In later years, Sommerfeld would not send his students to Bohr's

---

[12]"Ich habe Ihnen noch nicht zum Nobelpreise gratuliert; ich tue es jetzt, recht herzlich, auf die viertägigen Reise von Chicago nach Pasadena. ... Hoffentlich hilft Ihnen die allgemeine und frühe Anerkennung Ihres Werkes ... über die gelegentliche Misstimmungen hinweg... Ich will Ihnen nicht viel über Amerika erzählen, nur dass ich überall (fast überall) freundlich aufgenommen wurde. ... Ich will Ihnen auch nichts über das deutsche Elend vorklagen, obgleich ich mit gebundenen Händen aus der Ferne zuschauend es doppelt stark empfinde. Ich will Ihnen lieber von einigen wissenschaftlichen Erfahrungen reden. Gleich am ersten Tage in Madison sprach ich den jungen Van Vleck, ... über seine Berechnung des Bohr-Kemble'schen He-Modelles. Es kann nun wohl kein Zweifel sein, dass dies Modell und Ihre *fürchterliche* Vorstellung vom Magnetismus falsch ist... Ich trage hier überall über Ihre neuen Atommodelle vor, die ich durch farbige Diapositive erläutere. Nur Ihre K-Schale ersetze ich natürlich durch meine Ansicht vom He- Modell. Wenn Sie mir schreiben *wollen*, so adressieren Sie am besten an... Sie *brauchen* mir aber nicht zu schreiben. Freuen Sie sich Ihres jungen Ruhmes in einem wohlgeordneten Vaterlande und im Kreise einer lieben Familie! Und denken Sie daran, wie elend es uns Deutschen geht!" Sommerfeld to Bohr, 21 January 1923 (BCW 3: 502–4).

institute, and the two of them exchanged only a few formal notes on occasions of official jubilees.[13]

4.2 American Voyage

In the months following the announcement of the 1922 Nobel Prize, Bohr received a dozen letters from American colleges and universities with offers of permanent or visiting positions. In addition, Walter Colby from the University of Michigan, then on a visit to Europe, asked whether Bohr could recommend a promising young Scandinavian for an academic position in theoretical physics. With Bohr's and Kramers' recommendations, Klein obtained an appointment in Ann Arbor in the fall of 1923.[14] With regard to his own visit, Bohr replied that he could not accept invitations for the spring but was more willing to consider a short-term trip to America in the autumn of 1923. From a variety of tempting possibilities (Berkeley, Madison, Chicago, and Pittsburgh, among others), he chose to visit Yale and Amherst College. Neither of the two locations had a strong research program in physics at the time, and neither had a faculty member with any real connections to Bohr's work.

"[A college] of which I presume you have never heard,"—as the president of Amherst College in Massachusetts, Alexander Meiklejohn, introduced his institution to Bohr—offered him a newly established Simpson lectureship for delivering four lectures of general interest. Bohr's major presentation there, "The Atom," according to the printed invitation, was "intended particularly to meet the needs of teachers in secondary schools." It introduced the speaker as "the author of one of the two principal theories of atomic structure" (it is not clear what, in the mind of the organizers, counted as the second principal model—possibly the one by Gilbert N. Lewis in Berkeley). Silliman lectures at Yale University, according to the donor's will, were intended to be "such as will illustrate the presence and wisdom of God in the natural and moral world." University President J. R. Angell assured Bohr, however, that no explicitly religious statement was required. The foundation's money could be, and indeed already had been, used to bring over distinguished scientists to give "any course of lectures which was not positively and avowedly materialistic in its conception."[15]

Bohr's choice to accept these two invitations was apparently determined neither by the needs of scientific research and proselytizing, nor by substantial honorariums, but primarily by geography. His decision to "visit universities and institutions not

---

[13] See Sommerfeld's congratulations on Bohr's 60th birthday. In his acknowledgment of the latter, Bohr wrote: "it especially pleases me to think how the old, close scientific and personal connections between your circle in Munich and the one in Copenhagen are continuously perpetuated by the successful activities of the younger generation." Bohr to Sommerfeld, 23 October 1935.

[14] Colby to Bohr, 17 January 1923; 19 May 1923; Bohr to Colby, 29 January 1923; 8 June 1923.

[15] Meiklejohn to Bohr, 30 January 1923; J. R. Angell, president of Yale University to Bohr, 21 April 1923 (BGC).

too far from New York"[16] indicated the main purpose of his American trip. As he was leaving Copenhagen in September 1923, Bohr was not yet sure whether he would be able to meet with officials from the Rockefeller Foundation, but he wanted to stay in the proximity of their headquarters while awaiting their decision on his grant application. Preliminary contacts developed through the efforts of Hansen and Christen Lundsgaard, a Danish doctor who worked at the hospital of the Rockefeller Institute for Medical Research in New York. Hansen toured American spectroscopic institutions in the fall of 1922, supported by Bohr's recommendation letters. Soon after his visit, Lundsgaard contacted Bohr from New York:

> As you of course know, Dr. Hansen mentioned about two months ago in a letter to me that you were very interested in the possibility of obtaining economic support for your Institute… [T]he matter has been in my mind every time I was in situation of such a character or met people whom I thought I could interest successfully in it. Two days ago I had a chance to talk with Dr. Abraham Flexner who is very closely connected with the circles in this country from which money is sometimes obtained for international science and international education, and I found the occasion favorable… He realized fully, I think, that it would be an international calamity if these hard times should deprive you of sufficient money to run the Institute on an international scale… I think we must act quickly. Fortunately I meet Dr. Abraham Flexner often so I can push the matter through personal contact. It is needless to say, of course, that the whole thing including all your information, is considered strictly confidential. I think it would be nice if you would send reprints and pictures of the institute.[17]

Bohr was extremely lucky with the timing of his inquiry. Rockefeller philanthropists were just about to launch an ambitious program to fund European science, a shift from the foundation's main preoccupations with matters of public health and general education. The designer of that initiative, Wickliffe Rose, proposed establishing a new foundation within the Rockefeller system—the International Education Board (IEB)—and was preparing to depart on a tour to Europe to lay the groundwork for supporting academic research there. Not a scientist himself, but a historian and philosopher, Rose idealized the importance of pure science and of Europe as its irreplaceable source, with an inferiority complex then still common in America. The destruction by the Great War and subsequent economic crises threatened to undermine that source, which he wanted to preserve for the sake of the human race. Within the next few years, Rose's activities would make him, in Daniel Kevles' words, "virtually a central banker to the world science" (Kevles 1971b, 192). Rose designed IEB's elitist and internationalist strategy: to select a handful of European centers of excellence in several disciplines, provide institutional grants to make "the peaks higher," and distribute postdoctoral fellowships to scientists from other countries to travel to these institutions for advanced training and research. Postdoctoral fellowships had been a standard way to help educate new professors in America, but from his side of the Atlantic, Rose hardly reflected upon the fact that, for European scientists, this was a rather novel and relatively unfamiliar institution.

On 6 April 1923, upon receiving a cable from Copenhagen, Lundsgaard submitted a preliminary application on Bohr's behalf, arguing that "economic conditions in

---

[16]Bohr to Colby, 24 May 1923.

[17]Lundsgaard to Bohr, 16 March 1923 (BGC).

Denmark [were] such that it [was] becoming more and more difficult to obtain money for an institution of purely international scientific importance." While pushing the matter in the USA and exaggerating Danish economic troubles, which were not as serious as in some other European countries, Lundsgaard placed his own personal hopes on succeeding professor Hans Christian Gram as the medical chair in Copenhagen and the head of one of the two main university clinics there. At a meeting with Rose on 18 May, Lundsgaard stated that

> Dr. Bohr is now directing the work of from six to ten young scientific workers from various countries. He has not sufficient room and equipment even for these; he had to turn away a number of men because of lack of equipment; would like additional space and additional equipment. This would cost about $ 35,000. He thinks it possible that with $ 20,000 from abroad he could secure the remaining $ 15,000, and perhaps could get from the government the necessary increase of funds for maintenance.[18]

With assistance from Hansen, Lundsgaard completed a questionnaire for Rose about the Copenhagen institute, composed the grant application, and on 27 June submitted these documents to the Rockefeller headquarters. In his cover letter, Bohr emphasized "the peculiar character of the Institute" as providing

> close cooperation between theory and experiment… and this explains why many physicists from other countries, and not least from the United States, wish to study here. Naturally, I wish very much indeed to be able to receive as many properly qualified physicists as possible and to offer them good working conditions,

However, the available restricted space, equipment, and means "permit it to accept only a small number of applications." He also wrote about the need for additional instruments to extend experimental work on spectroscopy into the regions of infrared and X-ray radiation. During subsequent consultations, the requested amount was increased at the very last moment from $20,000 to $40,000. Bohr worried that this might be too risky a bid, but by Rockefeller standards, the change apparently was not that serious.[19]

Rose's letter proposing a personal meeting in New York crossed with Bohr's journey by sea from Denmark to America and did not reach him until late October 1923 in Amherst. Meanwhile, Hansen reported to Bohr that Lundsgaard had not performed very well during the interview in Copenhagen but that his appointment was still likely to succeed, which it did.[20] In recognition of his own important service in this business, Hansen was promoted in 1923, becoming the university's third ordinary professor in physics. Internal evaluations solicited by Rose from P. A. Levene, Rufus Cole, and Simon Flexner of the Rockefeller Institute assured that "as to Dr. Bohr there

---

[18]Lundsgaard to Flexner, 6 April 1923 (RAC. Projects. Denmark); W. Rose Diaries, 1922–23 (RAC. RF.12.1: 70–71); R. Pearce Diaries, 1923 (RAC. RF.12.1: 76).

[19]Rose to Lundsgaard, 19 May 1923; Bohr to IEB, 27 June 1923; Hansen "The Institute for Theoretical Physics of the University of Copenhagen"; Lundsgaard to Rose, 31 July 1923; Rose to Bohr, 30 August 1923 (the letter had not reached Bohr before his departure from Copenhagen and was sent again to Bohr at Amherst on 15 October) (RAC. Projects. Denmark).

[20]R. Pearce Diaries, 1923 (RAC. RF.12.1: 184. 21 November 1923); O. T. Avery to Rose, 16 October 1923 (RAC. Projects. Denmark); Hansen to Bohr, 1 October 1923.

could be no two opinions; his name would be contained in any list of outstanding physicists of the world." On 5 November, Rose conferred with Bohr in New York, and two weeks later the Rockefeller board approved the request of $40,000 for the enlargement and additional equipment of the Copenhagen Institute for Theoretical Physics.[21] Having written to Bohr about this decision and made the corresponding entry in his diary, Rose sailed off to Europe to select further sites of academic excellence worthy of Rockefeller's funding. Bohr returned to Copenhagen shortly before Christmas. Having seen a report about the Rockefeller grant in the *New York Times*, Pauli congratulated him on "the brilliant fulfillment of the main goal of your American trip," while expressing arrogant doubts about the capability of American audiences to understand the scientific subtlety of Bohr's theories.[22]

The grant to the Bohr institute was relatively modest in size, if compared to subsequent grants given by the IEB to other academic institutions, but it was the first of its kind, the earliest example of the Rockefeller philanthropic investment in European science. To become effective, it also required some matching contributions from local sources. In April 1924 Bohr informed Rose about an adjacent piece of land he had secured from Danish authorities, some additional funds for maintenance, and two new university positions, of a second scientific assistant and a second technician. The first installment of the Rockefeller funding became available in May, and the construction of the new building started in the institute's yard, which would last more than two years until completion. The so-called villa built on the premises of the institute would become Bohr's family house, while the third floor of the original building—initially used as the director's residence—provided office space for visiting researchers. Once the entire initial grant had been spent on construction, Bohr made a request to Rose for an additional $15,000 to purchase scientific instruments. Rose felt uncomfortable, and in the end Bohr received $10,000 from the Carlsberg Fond, and only a year later an additional $5,000 from the IEB.[23] Bohr's grant improved Denmark's chances of attracting the donors' continuing attention: August Krogh, Lundsgaard, and Hansen initiated and successfully applied for a much larger grant (eventually amounting to $300,000) from the Rockefeller Foundation for the construction of a physiological laboratories complex in Copenhagen under Krogh's direction, which also included a biophysical laboratory for Hansen.[24]

---

[21] W. Rose Diaries, 1922–23 (RAC. RF.12.1: 94–95, 98, 118, 128—Meetings with Bohr on 5 November 1923 and 21 November 1923); Rose to Bohr, 15 October 1923; Bohr to Rose, 30 October 1923; Rose to Bohr, 21 November 1923 (RAC. Projects. Denmark).

[22] "Americans eager to work with Bohr. Danish scientist may select students here under Rockefeller award," *New York Times*, 27 January 1924; "Sie den Hauptzweck Ihrer amerikanischen Reise so glänzend erreicht haben." Pauli to Bohr, 21 February 1924 (PWB, 147).

[23] Bohr to Rose, 16 April 1924; IEB to Bohr, 16 May 1924; Bohr to Rose, 29 November 1924; Rose to Bohr, 17 December 1924; Bohr to Rose, 8 January 1925; August Trowbridge Log, 25 September 1925; Bohr to Trowbridge, 6 November 1925, W. W. Brierley to Trowbridge, 9 December 1925 (RAC. Projects. Denmark).

[24] Krogh to G. Vincent, President of the Rockefeller Foundation, 16 April 1923; Rose to Pearce, 6 February 1924; Hansen to the Rockefeller Foundation, 1 July 1924; Lundsgaard to A. Flexner,

The Rockefeller funding arrived particularly timely for the development of quantum theory, just as Bohr's research program was entering an impasse. The construction of the new building would occupy most of his personal time and efforts during the subsequent two years and take Bohr away from pursuing his own research at the same pace as before. But, in accordance with Rose's schemes for IEB activities, the initial institutional grant was only the seed investment that had to be supplemented by additional "soft" funding for international fellows, primarily postdocs converging to that "center of excellence." Starting 1924, the lure of IEB postdoctoral fellowships substantially increased the traffic of foreign visitors to the Bohr institute and brought highly qualified young scientists, especially from Germany but also other large and small countries, to Copenhagen, along with new ideas, proposals, and directions for research. Bohr requested the first of such fellowships from the IEB already in 1924: for Werner Kuhn from Switzerland and Heisenberg from Germany, and also for Rosseland to go to the Mount Wilson Observatory in Pasadena. Transitory young researchers supported by the Rockefeller funds and other sources would soon help find new ways out of the conceptual crisis in quantum theory, culminating in the invention of quantum mechanics.

4.3   The Copenhagen Putsch

"The most interesting scientific experience that I learned in America is a study by Arthur Compton in St. Louis. From now on, the wave theory for X-rays would have to be dropped once and for all," declared Sommerfeld in his last substantive letter to Bohr in January 1923. In Germany, many subscriptions to foreign journals lapsed during the period of hyperinflation, making the spread of the news more difficult, but Sommerfeld popularized the Compton effect very widely upon his return from overseas in May 1923. The discovery quickly became the physics sensation of the year in Europe, recognized almost everywhere, except Copenhagen, as the definitive experimental confirmation of the existence of light quanta. Bohr remained practically alone among the major authorities of new physics as a steadfast resister to the idea of the corpuscular structure of light. Probably no other conceptual challenge troubled him more: Einstein's light quanta were "simply horrifying" to him, unreconcilable with phenomena of interference, and he continued struggling to preserve the classical wave theory of light to the very end and at heavy cost.[25]

Even Pauli became a defector. For most of 1923 he worked in Copenhagen as Bohr's private assistant in strict accordance with his director's agenda. He did not publish his 1922 calculations of the helium spectrum so that Kramers could complete his own. His research on multiplets followed Bohr's preferences in clear opposition

---

12 July 1924 (RAC. RF. 1.1.713, box 2, folder 19); Pearce to the Rector of the University of Copenhagen, 15 November 1924; Bohr to A. Flexner, 29 November 1924.

[25]Sommerfeld to Bohr, 21 January 1923; Bohr to Rutherford, 9 January 1924; Pauli to Bohr, 16 July 1923 (PWB, 102).

to those of Heisenberg and Sommerfeld. But on the issue of light quanta, Pauli could not keep silent, however, European academic habitus of the time gave professors the ultimate responsibility for research by junior scholars and subordinates. In that hierarchical, pre-peer review era, directors provided quality control and had to approve all papers originating from their institutes before they could be sent to a journal for publication. Pauli knew that Bohr would not agree with a paper on light quanta and, in the end, resorted to a little (or not so little) insubordination or hooliganism. While away from Copenhagen on a short summer vacation, he returned to Hamburg and wrote a paper on the Compton Effect there, even though it still, officially, carried the institutional affiliation "in Copenhagen" (Pauli 1923).

Feeling increasingly under assault, Bohr made his last desperate attempt to save the wave theory of light after he returned from the USA in December 1923. This time, he worked with an American postdoctoral fellow, John Clarke Slater. American research students had been seen frequently at European universities since the second half of the nineteenth century and especially after 1919, following the launch of a major program of postdoctoral fellowships by the National Research Council (Kevles 1971a; Assmus 1993). The Copenhagen institute often received visitors supported by a smaller philanthropic institution, the American-Scandinavian Foundation, which had its own program of academic exchanges and granted approximately 20 post- or pre-doctoral fellowships annually to Americans who wanted to study in Sweden, Denmark, or Norway. Many of the recipients were of Scandinavian descent, and in that early, pre-IEB period, they came to the Bohr institute at an average rate of one per year: Bruce Lindsay, Frank Hoyt, Harold Urey, Anton Udden, and Eric Jette. Academically, not all of them were sufficiently prepared or knew exactly what Bohr's research was about. Since they arrived with their own funds, Bohr did not need to provide much and did not expect much from them, either (Davies 1985).

This partly explains the frustration of Slater, a young PhD from Harvard who arrived in Copenhagen by the end of 1923. Politically conservative, traditionally racist and sexist, Slater was used to the Harvard-style prestige and entitlement, which was not automatically extended to him in Europe.[26] Bohr's exchanges with Pauli that year show that they still looked down upon American theoretical physics with the typical sense of European superiority. In the case of Slater, such a dismissive attitude was completely unjustified: He arrived well versed in quantum theory, and his choice of Copenhagen was very conscious. Moreover, he already had a publication to his credit and brought with him an original idea of his own, a complex way to combine light quanta and the virtual wave field. This time it was Bohr who was not quite prepared for such a surprise, which left Slater with the disappointing feeling of

---

[26] Slater on little pleasures of graduate student life at Harvard: "The food was quite good, the prices very reasonable, and the service was by colored waiters, very much the old family retainer type. One of the traditions was that no women might enter the room. There was a gallery at one end, and once in a while some one would bring some girls onto the gallery to show them the sights. As soon as the students would catch sight of the girls, they would start shouting, banging with their spoons on their glasses, and throwing hard rolls up at the targets, until the girls fled… I was glad to have had a chance to see this relic of the old days." *A Physicist of the Lucky Generation. Autobiography,* 215 (Slater Papers, APS).

having been treated by Bohr and Kramers "as an assistant, to work on their ideas." Unfortunately for Slater, in Copenhagen light quanta were still regarded as heresy. He helplessly watched as his original proposal transformed through discussion between Bohr and Kramers into something quite different from what he had intended: The idea of virtual radiation field emitted by the atom in a stationary state was preserved, but light quanta were eliminated from the picture entirely. Bohr and Kramers could achieve this elimination only by sacrificing a previously sacred physical principle. In a risky move, they proposed to abandon the strict validity of the laws of conservation so that energy and momentum were preserved only in the sum total of a great many atomic processes, but not in each individual microscopic interaction between light and matter.

Revolutionary minds in the new physics grew used to sacrificing many of the fundamental classical concepts and principles, but Bohr's proposed deviation from strict energy conservation appeared to many of them as too destructive and going too far. Slater, a young novice, apparently also disagreed with it, but did not feel empowered to resist the authority of his coauthors: "Bohr is a very open minded and fair-minded person; just the same, there is of course more or less that one must (feel) think as he does, and I am unable to tell whether this would interfere with my happiness or not," he shared his doubts in a letter to his father (Schweber 1990, 354). The resulting paper proposing the famous (or infamous) Bohr-Kramers-Slater theory (BKS) was ready in January and published in April 1924. Bohr sent the German version of the manuscript to Pauli for linguistic editing, but preemptively refused to hear objections to its content before the publication: "if there are disagreements on this point, we can always quarrel [later] in April."[27]

Of course, there were disagreements, but when Pauli visited Copenhagen during Easter, he, too, could not resist Bohr's argumentation or authority: "[You] succeeded at that time in silencing my scientific consciousness, which revolted strongly against this interpretation. But this was the case only for a short time," he confessed to Bohr half a year later. Once Pauli left Copenhagen, he was once again "a physicist totally opposed to this interpretation of the radiation phenomena."[28] Opposition to the BKS proposal was shared widely, with only a few notable exceptions, such as Schrödinger in Zürich. In Copenhagen, however, Bohr and Kramers, assisted by a new IEB postdoctoral fellow from Germany, Heisenberg, continued for nearly a year to write papers in line with what they called the "true faith," and what Pauli called the "Copenhagen Putsch."[29] Yet even within this narrow circle, a tacit subversion took place. After the attempt failed, it became clear that both Kramers, in his theory of dispersion, and Heisenberg, in his theory of resonance fluorescence, had been careful enough to use only the language of virtual oscillators without relying on the

---

[27] (Bohr, Kramers, Slater 1924); Bohr to Pauli, 16 February 1924 (PWB, 146).

[28] "Es ist Ihnen damals gelungen, mein wissenschaftliches Gewissen, das sich gegen diese Auffassung stark auflehnte, durch Ihre Argumente zum Schweigen zu bringen. Dies war aber nur kurze Zeit der Fall und... stehe ich heute dieser Auffassung der Strahlungserscheinungen als Physiker vollkommen ablehnend gegenüber." Pauli to Bohr, 2 October 1924 (PWB, 163), English translation from (BCW 5: 32).

[29] Bohr to Kramers, 23 September 1924; Pauli to Kramers, 27 July 1925 (BCW 5: 52, 88).

riskiest part of the proposal, the violation of conservation laws. In retrospect, Bohr appears to be almost alone in sincerely abandoning the strict conservation of energy and momentum.

In Göttingen, Franck found himself in a difficult situation, wavering between his loyalty to Bohr and his conscience as an experimental physicist whose lifelong research relied on conservation laws: "I am ashamed to admit that, with regard to your view of the statistical validity of the principle of energy conservation, I, too, am in a state of statistical equilibrium." When communicating to Bohr his most recent results on the Ramsauer effect, Franck apologized that he had allowed himself "to relate this somewhat divergent opinion at the end of the work."[30] Once the adventure was over, and Bohr had acknowledged the fiasco, Franck consoled him: "Your revolutionary putsch was certainly quite an excellent thing, which had an uncommonly stimulating effect... Personally, however, I am glad." The word "putsch" came directly from Germany's political realities of the time, with its rightist and leftist revolts. Which particular political label to assign to the latest revolt in physics reflected the commentator's prejudices. Bohr (and Franck) called it a "revolutionary effort," while Pauli blasted it as "reactionary," rejoicing about its failure and the return to "positive progress" in a letter to Kramers: "I now feel less lonely than about half a year ago when (spiritually and spatially) I found myself rather alone between the Scylla of the number-mystical Munich school and the Charybdis of the reactionary Copenhagen Putsch, propagandized by you to fanatical excesses!"[31]

During that year, Born strove to become as close to Bohr as possible. Even prior to reading the BKS paper, he congratulated Bohr on its success: "Although I have only Heisenberg's short oral report, I am quite convinced that your new theory is correct" and agreed that his assistant Heisenberg should go to Bohr for a year: "I will certainly miss him... but his interest is more important than my own, and your wish is crucial to me." (Born was planning to spend part of that year in America anyway).[32] Born placed his own hopes of resolving the difficulties in quantum theory on a "truly discontinuous quantum mechanics" in which all continuous physical quantities would be replaced by discrete sets. He added the language of virtual oscillators to this proposal as a friendly gesture toward the Copenhagen "true faith," but, just like

---

[30]"Ich schäme mich zu gestehen, daß ich in Bezug auf Ihre Auffassung der statistischen Gültigkeit des Energieprinzips auch mich in einem statistischen Gleichgewicht befinde." "diese etwas abwe-ichende Meinung am Schlusse der Arbeit zu schildern." "Ihr revolutionärer Putsch war doch eine ganz grossartige Sache, die ungemein anregend gewirkt hat. ... Personlich bin ich allerdings froh." Franck to Bohr, 5 March 1925; 20 April 1925 (BCW 5: 345, 351).

[31]Franck to Bohr, 24 April 1925; Bohr to Franck, 21 April 1925; Bohr to Fowler, 21 April 1925. "So fühle ich mich denn jetzt weniger einsam als etwa vor einem halben Jahr, wo ich mich (geistlich wie räumlich) zwischen der Scylla der zahlenmystischen Münchener Schule und der Charybdis des von Ihnen mit zelotischen Exzessen propagierten reaktionären Kopenhagener Putsches ziemlich allein befand!" Pauli to Kramers, 27 July 1925, English translation in (BCW 5: 88).

[32]"Obwohl ich nur den kurzen mündlichen Bericht Heisenbergs habe, bin ich ganz überzeugt, daß Ihre neue Theorie das richtige trifft," "Ich werde ihn natürlich vermissen, ... aber sein Interesse geht vor dem Meinen, und Ihr Wunsch ist mir ausschlaggebend." Born to Bohr, 16 April 1924 (BCW 5: 299). See also Born to Heisenberg, 26 November 1924; Bohr to Born, 1 December 1924; 9 December 1924; Born to Bohr, 6 December 1924.

Kramers and Heisenberg, did not actually rely on the non-conservation of energy. In the fall of 1924, Born sent a manuscript on the chemical bond he had written together with Franck to Copenhagen and was prepared to revise it in order to accommodate Bohr's latest views. As a recognition of their special relationship, Bohr invited Born to Copenhagen in February 1925 for a public lecture (on the quantum theory of molecules) and informal discussions (on the new theory of radiation).[33]

By that time Born had already indicated that in the personal split between the two main authorities in quantum theory, he had taken Bohr's side against Sommerfeld's:

Today I got a letter from Sommerfeld in which he, somewhat patronizingly, expresses his appreciation for my atomic mechanics book, which I sent him, but also harsh criticism. He starts by saying: "Your book is, as I expected, quite Bohr-devout." Then come complaints about the lack of citations of Sommerfeld's discoveries (whereby he is partially correct) and about the omission of Sommerfeld's theories; e.g. he reproaches me that I did not discuss the He model... he designed, and that I use your new conception of radiation too much. Heisenberg, however, finds that I am not using the latter consistently enough. I am thus caught in a difficult position between Munich and Copenhagen and must figure out how to live on. Overall, my philosophically sensitive head pulls me towards the North.[34]

Meanwhile in Berlin, Walther Bothe and Hans Geiger conducted an experiment that refuted the BKS proposal by demonstrating that energy and momentum simultaneously conserved in an individual act of scattering between a light quantum and an electron (Fick and Kant 2008). Born learned of their preliminary results while visiting Berlin in January 1925 on the eve of his Copenhagen trip, but was still "not at all convinced" and "had objections." He changed his opinion by April. Once Bohr himself admitted the failure after receiving a letter from Geiger, Born, too, quickly acknowledged that "I, too... had come to consider this theory as impossible" and sent Bohr an already finished manuscript of a theoretical calculation in which light quanta were guided by virtual radiation waves. Bohr did not give his approval for publication. Having put his reputation on the line and suffered a damaging blow, he saw the only remaining tactics in "relegating our revolutionary attempt into oblivion, as painlessly as possible." After the defeat of his last stand against light quanta, Bohr accepted Pauli's apophatic view that "it is not possible to describe physical processes

---

[33]Born to Bohr, 15 December 1924; 15 January 1925 (BCW 5: 301–3).

[34]"Heute bekam ich einen Brief von Sommerfeld, in dem er mir, etwas von oben herab, seine Anerkennung für mein Buch *Atommechanik* ausspricht, das ich ihm geschickt hatte, zugleich aber scharfe Kritik übt. Er fängt gleich so an: "Ihr Buch ist, wie ich es erwartet habe, ziemlich Bohr-fromm." Dann kommen Klagen über fehlende Zitate Sommerfeldscher Entdeckungen (wobei er zum Teil recht hat) und über Auslassung Sommerfeldscher Theorien; z.B. hält er mir vor, daß ich das von ihm entworfene, drehimpuls-freie He-Modell nicht diskutiert habe, und daß ich zu sehr Ihre neue Strahlungsvorstellung benütze. Heisenberg hat umgekehrt gefunden, daß ich diese nicht konsequent genug benütze. So sitze ich jammervoll zwischen München und Kopenhagen und muss sehen, wie ich weiterleben kann. Im ganzen zieht mein philosophisch angekränkelter Kopf mich nach Norden." Born to Bohr, 15 January 1925 (BCW 5: 304).

with the help of simple visual images,"[35] which in a way was a repudiation of his program based on the 1913 atomic model.

His sense of a crisis was quite intense. It was one thing to be prepared to abandon some postulates of the long existing, classical physics, but quite a different matter to accept the failure of one's own brainchild, the revolutionary quantum theory of the atom, only ten years old, and just recently so promising and opening breathtaking new perspectives. In hindsight, one can probably say that the so-called old quantum theory still had plenty of potential for development and for solving new problems. Sommerfeld thought so, as did many others in the bourgeoning field of quantum spectroscopy. Many of the existing spectroscopic difficulties, as is turned out, would be explained by the idea of electron spin (a visual model), which could have relieved much pessimism and given the theory a second breath. But the spin did not come to the rescue before it was too late: it was suggested in late 1925 independently from, but just after the more radical quantum mechanics. By early 1925, Bohr felt that his program had lost momentum, and similar feelings were shared within the small circle between Copenhagen and Göttingen, mostly in letters and private communications, rather than publicly. The idea of a much more general, open crisis, affecting the entire community, is a later extrapolation and post hoc rationalization following the spectacular success story of quantum mechanics. As we will see in the next chapter, that success would come primarily from postdocs who managed to liberate themselves, at least partially, from the control of established professors in the field and from their exhausted agendas.

---

[35]"ich selbst... ebenfalls dazu gekommen war, diese Theorie für unmöglich zu halten," "unseren Revolutionsversuch möglichst schmerzlos in Vergessenheit zu bringen." "eine einfache Beschrei-bungsmöglichkeit des physikalischen Geschehens mittels anschaulicher Bilder ausschliesst." Bohr to Franck, 21 April 1925; Born to Bohr, 25 April 1925; Bohr to Born, 1 May 1925.

Chapter 5
Revolt of the Postdoc

5.1 Marginal Intruders and the Light Quantum

The hypothesis of light quantum, initially proposed by Einstein in 1905, for a dozen years struggled to gather enough support. Circa 1900 Lorentz managed to incorporate the electron—a discontinuous atom of electricity—into the previously fully continuous Maxwellian theory of the electromagnetic field. The resulting theory, currently known as classical electrodynamics, combined field-like and particle-like concepts in a unified mathematical scheme. Encouraged by this success, Einstein hoped to replicate a similar feat, however, the case of optics and light resisted all attempts by him and others to achieve a theoretical synthesis and resolve contradictions between corpuscles and waves within a single consistent model (Kojevnikov 2002). Many experiments, typically those related to the absorption and emission of light by matter, could be easier understood when light was imagined as having a quantum structure. But this discontinuous model contradicted other, no less numerous optical phenomena, typically related to the propagation of light through space, including interference and diffraction, which had been fully explained by the wave theory of light and verified in most precise measurements. At the first Solvay conference in 1911, two dozen international experts worked out a compromise, a preliminary consensus. They recognized the concept of discontinuous quanta, despite some objections, as a fundamental thing and major challenge to the existing foundations of physics. But the source of those mysterious quanta, with reasonable caution, was assumed to exist somewhere in the complex and still unexplored problem of interaction between matter and radiation, rather than in the structure of light itself, which for the time being remained understood as a continuous wave. Einstein, who after a struggling academic career had just been freshly minted as a professor in Prague, reluctantly accepted the wisdom of his senior peers, or at least stopped insisting on his hypothesis. When Einstein was elected to the Prussian Academy of Sciences in 1913, Planck as the nominator mentioned the light quantum as his only serious

A. Kojevnikov, *The Copenhagen Network*, SpringerBriefs in History of Science and Technology, https://doi.org/10.1007/978-3-030-59188-5_5

blunder, excusable against the background of his other great accomplishments in physics.[1]

Intellectual fortunes of the light quantum reversed dramatically after 1918, as the concept rapidly gained popularity among the younger, postwar generation of physicists. The inspiration, or frequent common reference, typically came from Einstein's 1916 theory in which he introduced the coefficients of spontaneous/induced emission/absorption of radiation.[2] Einstein formulated his new results in a manner that naturally allowed, almost provoked, the imagery of light quanta. He stopped short of openly using that language himself, but many readers of his article were much less cautious and readily made that step. Their backgrounds and approaches were noticeably different from those of the older, prewar generation. Not as thoroughly trained in classical thermodynamics, statistics, and electromagnetism, they lowered the bar for discussable hypotheses and could more easily generate and propose ideas that looked either too weird or trivially wrong (or both) from the point of view of established theories. They did not hesitate to discuss light quanta in a more straightforward, realistic manner, as basic, fundamental particles, rather than as a correction to the existing model of light. More tolerant, or less cognizant, of the unresolved theoretical contradictions that continued to bother Einstein and undermined his desire to push the light quantum idea further, they sometimes did not fully comprehend all the radical, unintended consequences of their own conceptual innovations.

A great many ideas generated in such a way were, and remained, amateurish and marginal, but a few did manage, against the odds, to make a huge impact on physics at the time. But even in these success stories, the experiences of young scientists were not as rewarding. The authors often came from the margins themselves—where unusual proposals had fewer chances of being killed at birth by professional criticism. They had a hard time bringing their unconventional ideas to the attention of recognized experts, and even when that happened, and when the credit and citations to the original author were properly given, they still often lost control of further developments, seeing their dearest thoughts adopted, taken away, transformed, reconceptualized, or plainly abused by more authoritative scholars. The important novelties, in this sense, were separated from the life trajectories of their authors, who, even having achieved name recognition, often remained where they came from originally, on the margins of the disciplinary community.

The case of Slater discussed in the preceding chapter partially fits this description, but an even more appropriate case in point is that of the young Indian physicist, Satyendra Nath Bose. Bose was educated in British India into the emerging group of

---

[1] "Daß er in seinen Spekulationen gelegentlich auch einmal über das Ziel hinausgeschossen haben mag, wie z. B. in seiner Hypothese der Lichtquanten, wird man ihm nicht allzu schwer anrechnen dürfen; denn ohne einmal ein Risiko zu wagen, läßt sich auch in der exaktesten Naturwissenschaft keine wirkliche Neuerung einführen." (Planck 1975). For a detailed history of the concept of photon, see Hentschel (2018).

[2] Einstein (1916), Small (1986). Indirectly debating, Einstein and Bohr several times borrowed and reinterpreted each other's ideas in such a way, as to suit their otherwise opposing views on the nature of radiation. Thus Einstein's 1916 theory relied on Bohr's concept of discrete atomic states and, in return, inspired Bohr's correspondence principle.

Bhadraloks—indigenous intelligentsia who received European-style training and a modicum of European lifestyle from colonial authorities. The latter aimed to create a local version of the service class to help run the bureaucratic colonial machine and were partially successful in this goal. But at the same time, it was also the *Bhadraloks* who shaped the emerging ideas of modern Indian nationalism, combining elements of native tradition with values of science and modernity and giving voice to the growing anti-colonial sentiment. After receiving his M.Sc. in mixed mathematics at the University of Calcutta in 1915, Bose became a lecturer in physics there and in 1923 took up a professorship at Dacca University. During World War I, his attachment to science and to anti-colonial nationalism combined into his growing obsession with relativity and the quanta as German, and therefore anti-British, concepts. He could read German and found some, albeit limited, access to literature on quantum theory in the personal library of another teacher who originally came from Austria. Bose's paper of 1923 dealt with a basic, textbook topic, the derivation of the Planck law, yet attempted to follow a new and strictly corpuscular road to it, without any use of wave-like concepts (Banerjee 2016).

He managed this at a cost of introducing, unintentionally, a different kind of statistics, now called the Bose-Einstein statistics, which Ehrenfest helped to clarify as describing indistinguishable particles.[3] "I was not a statistician to the extent of really knowing that I was doing something totally different from what Boltzmann would have done, from Boltzmann statistics," recalled Bose years later. He submitted the paper to the *Philosophical Magazine* but received no reply. A year later, in 1924, Bose sent his manuscript directly to Einstein, who became inspired, translated it into German, and ensured swift publication. Einstein quickly adapted Bose's approach to develop his own theory of the ideal quantum gas, which guaranteed attention and fame (Bose 1924; Einstein 1924). But when later that year Bose travelled to Europe himself—Paris and then Berlin—on a scholarship, he felt marginalized and alienated, in part because of his personal shyness, but also from the arrogant, patronizing attitudes of European scientists. Even Einstein, who was not exactly a model of social empathy, was critical, in print, about Bose's second paper. "Heartbroken, Bose returned to India in 1926, concentrating on teaching and guiding students" (Banerjee 2016, 174).

On the opposite pole of social and racial hierarchies, yet also marginalized in the discipline, stood Louis, 7th duc de Broglie, a scion of one of France's most aristocratic families. He and his elder brother Maurice became professional physicists by defying strong pressure from relatives, who saw their forfeiture of military in favor of scientific careers as an affront to noble reputation. Maurice's experimental research on X-ray spectroscopy and absorption made him sympathetic toward the idea of light quanta. A house theoretician in his brother's private lab, Louis published a series of papers in 1922–1924, pushing the analogy between matter and radiation

---

[3]Ehrenfest was unlucky in the sense that he understood the discrepancy between the statistics of independent quanta and that of indistinguishable quanta better than either Einstein or Bose, a complication that deterred him from making this step in his own work, but, on the other hand, allowed him to point out the difficulty to Einstein in 1924.

much further, first into an assumption that quanta of light have small mass, and then into the reverse—that every existing material particle comes with an associated wave of characteristic length. French theoretical physicists of the time had mastered Einstein's relativity, but mostly ignored the German quantum theory. De Broglie happened to read German and knew some of the German-language literature on the topic, but there was hardly anyone else in France prepared to understand his theoretical ideas. Luckily, one of his mentors, Paul Langevin, was a personal friend of Einstein, and at a meeting in neutral Geneva in the summer of 1924 told Einstein about Louis's forthcoming doctoral defense at la Sorbonne (Darrigol 1993).

Einstein received a copy of the thesis before the end of the year, just in time to include a welcoming citation in his second communication on the quantum theory of ideal gas (Broglie 1924; Einstein 1925). Einstein, too, was a strong proponent of the analogy between matter and radiation, if not of the wave–particle duality (Kojevnikov 2002). His footnote caught the attention of Schrödinger in neutral Switzerland, who was able to get a copy of de Broglie's *These* in November 1925 and during the Christmas holidays transformed the Frenchman's concept of matter waves into one of the versions of the new quantum theory, wave mechanics. "He retained only the waves and forgot about the particles "sliding" on the waves. The result was the Schrödinger equation. In a single transformation, Louis de Broglie's idea was betrayed and glorified," concludes Olivier Darrigol (1993, 359). De Broglie's subsequent efforts to participate in the development of quantum mechanics were effectively sidelined by the German-centered disciplinary community, and after a negative reception at the 1927 Solvay conference, he abandoned attempts to further develop his own original interpretation of the new theory (Cushing 1994; Bricmont 2016, 264–69).

5.2 Inflation and Quasi-Free Postdocs

Between 1870 and 1920, German students did not need to, and as a rule did not, go abroad for advanced study. Confident in the superiority of the training they provided, German professors welcomed students from other countries rather than sending their own away to seek foreign expertise. A successful path toward an academic career in Germany would typically start with an assistantship, private apprenticeship (the position of "graduate student" did not formally exist), and a doctoral dissertation under one of the important professors in the discipline. Subsequent stages usually included several moves between universities within the German-speaking world (including Austro-Hungary and Switzerland), while the young scholar rose through the academic ranks with *Habilitation* to *Privatdozent*, and later, with some luck, to the position of extraordinary and ordinary professor. Assistants, both private and university-employed, worked for the professor in teaching and research. *Privatdozenten* were supposed to be independent scholars with the right to teach their own courses and collect fees from attending students, but they still depended upon institute directors for access to resources, equipment, and in other respects.[4]

A particularly successful professor could eventually become the head of his own "school" or "seminar," an informal but usually state-funded institution to train PhDs for the entire discipline, a forerunner of the American "graduate school." Sommerfeld ran such a seminar in Munich from 1906 on, designed to be the "plant nursery" (*Pflanzstätte*) for theoretical physics. Managing the production of PhDs gave the professor a powerful voice in recommending candidates for appointments and promotion. By 1928, Sommerfeld's former pupils occupied about one third of all ordinary professorships in theoretical physics at German-speaking universities and engineering schools. Educational ministries of various German lands often approached Sommerfeld with requests to recommend or evaluate candidates for professorial chairs in the field. In his rankings of available scholars, he typically surveyed the entire pool, not only his own students. A major force, together with Planck, behind many academic appointments, Sommerfeld acquired in Germany the reputation of the *éminence grise* of theoretical physics.[5]

Professors usually controlled the traffic in assistants, as well as the selection of candidates for *Habilitation*. In such a manner, Sommerfeld loaned his student Pauli to Born as an assistant for the 1921/22 year, whereby the two professors made deals and arrangements about the careers of their protégés. In January 1923, Born asked Sommerfeld's permission to hire Heisenberg:

> I would like to have a *Privatdozent*, for I am too burdened with lecturing. Paul Hertz does not count because he has changed over to philosophy, and my doctoral candidates... have not proceeded far enough; also they cannot be compared to Heisenberg. You have Wentzel, and I assume that Pauli will return to you after a year. Could you under these circumstances give up Heisenberg and persuade him to get his *Habilitation* in Göttingen? I shall, of course, make sure that he will be well off financially... Naturally, I would also welcome Pauli very much; but he cannot stand, as he claims, the life in a small town.[6]

At the time of writing, Heisenberg was spending a semester with Born while Sommerfeld travelled in America. In summer 1923, he defended his PhD in Munich, and then, with Sommerfeld's approval, moved to Göttingen as Born's private assistant. Pauli's and Heisenberg's subsequent career paths, however, did not materialize exactly as envisioned by their supervisors: Pauli did not return to Munich, while Heisenberg did not remain in Göttingen for any significant time after his *Habilitation* lecture in June 1924. The main factor that shattered the social fabric of the German academe and professors' control over the careers of junior apprentices was hyperinflation.

Its first effects were already noticeable in the summer of 1922, and by September, because of the "latest crash of the Mark," Pauli was unable to even purchase a railway

---

[4]For a detailed explanation (for outsiders) of the established German academic system and its career patterns, see Forman (1967, 59–122). Occasional German students did study abroad after receiving their PhDs, for example Hahn and Geiger. I am thankful to Dieter Hoffmann for this hint.

[5]See Sommerfeld (1984, 111–17) for the name list of fifty of Sommerfeld's students who embarked upon academic careers in Germany and in other countries.

[6]Born to Sommerfeld, 5 January 1923, English translation from HDQT 2 (73).

ticket from Hamburg to Copenhagen to start his temporary appointment there and had
to ask Bohr to send him an advance in Danish crowns.[7] Schrödinger, who had just
recently obtained a professorship in financially secure Switzerland, congratulated
Pauli on his escape to a neutral and safe land: "I am very pleased that you are going
to Bohr... It will be better for you there. I hope you will not come back in any
foreseeable future, for I feel dreadful when I imagine having to go back to Germany,
and after a year you will feel the same."[8] They understood each other as fellow
Austrians: Both knew how science in their homeland was decimated immediately
after World War I, making it impossible to live on academic salaries and driving many
scholars to emigrate. Both had then left Austria to seek better career opportunities in
the still relatively stable economy of Germany. A few years later when the financial
collapse had caught up with Weimar Germany, both panicked on the grounds of their
earlier experiences. Returning to Hamburg the following summer, Pauli was relieved
to discover that his worst fears did not materialize:

> When I left Copenhagen, I feared that by now the economic and political conditions in
> Germany would have a paralyzing effect on scientific work in the institutes. And I was then
> very glad to see that—at least in Hamburg—this is not the case at all. Now that Stern is also
> there, they are enjoying a lively academic life.[9]

Indeed, inflation did not damage science in Germany as severely as in the much
smaller splinters from former Austro-Hungary. German research and publishing
continued as actively as before, even if a prohibitive exchange rate prevented
subscriptions to foreign publications and undermined opportunities for foreign travel.
The infrastructure for research—institutes and laboratories, built and equipped during
the imperial period before the Great War—was still far better than anywhere else in
Europe. Professors and other salaried academics maintained livable incomes adjusted
for inflation, while grants from the emergency fund *Notgemeinschaft der Deutschen
Wissenschaft* partially compensated for losses in research support. The negative
effects of German hyperinflation can be described as structural rather than outright
destructive. Arguably the most difficult fate befell younger academics who, like Pauli,
were caught between the doctorate and their first professorial appointment. In the
past, at this stage in their academic careers, young scholars would have held positions
at universities as independent *Privatdozenten*, but inflation made this class almost
totally extinct, as it was no longer possible to sustain one's livelihood on middle class
savings or on "soft money," such as student fees. In response to the crisis, German

---

[7]Pauli to Bohr, 5 September 1922, and Bohr to Pauli, 8 September 1922, both in (PWB, 62–64).

[8]"Freue mich sehr, daß Sie zu Bohr kommen.... Es wird Ihnen dort besser gehen. Ich hoffe, Sie
kommen in absehbarer Zeit nicht von dort zurück, denn wenn ich mir denke, daß ich jetzt wieder nach
Deutschland sollte, so graut mir, und so wird es Ihnen nach einem Jahr auch ergehen." Schrödinger
to Pauli, 8 November 1922 (PWB, 69).

[9]"Als ich von Kopenhagen abreiste, habe ich gefürchtet, in Deutschland würden die wirtschaftlichen
und politischen Verhältnisse inzwischen auf den wissenschaftlichen Betrieb in den Instituten
lämend gewirkt haben. Und ich war dann sehr erfreut zu sehen, daß das—wenigstens in
Hamburg—keineswegs der Fall ist. Seitdem auch noch Stern dort ist, herrscht dort ein sehr reges
wissenschaftliches Leben." Pauli to Bohr, 16 July 1923 (PWB, 102).

professors increased the number of positions for assistants and created special assistantships for *Privatdozenten*, to the effect that, at a practical level, the latter no longer constituted any significant career improvement over the former.[10]

Had Sommerfeld invited Pauli to become a *Privatdozent* in Munich in the summer of 1922, Pauli would have felt honored and lucky. But in the meantime, hyperinflation had hit and undermined the resources of German professors, sending Sommerfeld, and later also Born, on well-compensated lecture tours to America. One year later, after Sommerfeld's return, Pauli declined the belated, if tempting, offer from his teacher and doctoral adviser:

> It is very friendly of you that you wish I should finally receive my *Habilitation* in Munich. Yet this is a difficult matter. On the one hand, colleagues in Hamburg are urging me to do my *Habilitation* there... On the other hand, Bohr would like me to come back to him after he returns from America.

"For a variety of reasons," apparently both scientific and financial, Pauli hoped to spend some additional time in Copenhagen, should Bohr again need his help with writing papers in German. Pauli was perfectly aware that Denmark could not offer him anything in the long term, but as a temporary shelter in unstable times, an assistantship in Copenhagen looked more attractive than a German *Privatdozent*:

> I will certainly not be able to stay here [in Copenhagen] forever and must sooner or later habilitate at one of the German universities... Bohr's [possible] offer, however, makes me inclined to leave the question of my *Habilitation* open, for now... The only thing certain is that I will still spend the coming semester in Hamburg... *What happens later I know as little as an electron knows in advance where it will jump in* 10^{-8} *s* (I have only described the forces deflecting me from Munich, but... of course, very strongly attractive forces come from Munich as well).[11]

The emphasized sentence is quite provocative, as Pauli compared his personal feelings of uncertainty and professional anxiety to that of a quantum particle. At the moment of writing, Pauli's professional future looked very insecure indeed. Formerly, the career paths of younger academics in Germany resembled the trajectories of classical particles, determined by the external forces of their professors. Due to hyperinflation, however, those trajectories now had to run through the region of

---

[10]For more on the *Notgemeinschaft* and a comparison of German and Austrian inflation and their effects on academic life, see Forman (1967, 206–37). For observers' remarks on the disappearance of *Privatdozenten*, see Assmus (1993, 178).

[11]"Es ist sehr freundlich von Ihnen, wenn Sie wünschen, daß ich mich schließlich in München habilitieren soll. Nun ist es damit eine sehr schwierige Sache. Einerseits drängen die Hamburger sehr, daß ich mich dort habilitieren soll... Anderseits will Bohr gerne, daß ich wieder zu ihm komme, sobald er von Amerika zurückkehrt. Dies wäre mir natürlich aus vielen Gründen sehr gelegen, aber es hat auch seine Schwierigkeiten. Denn ich werde ja doch nicht immer hier bleiben können und mich früher oder später an einer deutschen Universität habilitieren müssen... Bohrs Angebot bestimmt mich aber doch zunächst, die Frage meiner Habilitation noch offen zu lassen... *Was später geschiet, weiss ich vorläufig ebensowenig, wie ein Elektron weiss, wohin es nach* 10^{-8} *s springen wird.* (Ich habe nur die von München ablenkenden Kräfte beschrieben, doch... selbstverständlich auch sehr starke anziehende Kräfte von München ausgehen)." Pauli to Sommerfeld, 6 June 1923 (PWB, 94, emphasis added).

negative finances, *Privatdozentur.* One could still try to live on an assistant's salary and essentially perform the duties of an assistant, or, like Pauli, abandon the notion of traditional trajectories altogether, become quasi-independent and explore new paths via quantum jumps through transitional metastable states, such as postdoctoral fellowships or temporary positions abroad. This quantum option was, of course, less predictable than the classical one: Pauli's hope to get another invitation from Bohr for a longer-term appointment in Denmark did not materialize (possibly because he showed himself unreliable, in Bohr's eyes, on the issue of light quanta and on the BKS theory). In February 1924 Pauli received his *Habilitation* and became a *Privatdozent* in Hamburg, from where he would occasionally travel to Copenhagen for short visits and discussions with Bohr.[12] His metaphor, however, works both ways: Pauli's (and Heisenberg's) personal insecurities and perceptions of the uncertain social world could also affect their thoughts and intuitions regarding the strange behavior of electrons and atoms, contributing, first, to their rejection of predictable trajectories for microscopic objects and then to further ideas regarding acausality and quantum uncertainty.

Postdoctoral positions offered a tempting way to survive the temporarily unsustainable stage in an academic career. Traditionally, German and, more generally, European academic culture lacked the notion and the institution of such fellowships. It first developed overseas, in late nineteenth-century USA, as a palliative measure aimed at compensating for the perceived inferiority of American PhDs in comparison with European ones, and quickly gained popularity as a method of providing advanced training for future college professors in the USA. By the turn of the century, aspiring young American academics, having completed their PhD at home, would typically spend a year or two visiting European universities supported by either private means or a fellowship. Postdoctoral study became so firmly accepted in the USA that it was taken for granted as the default and most appropriate way of promoting research-oriented expertise. The Rockefeller officials possibly did not fully understand that by offering it to Europeans, they were importing their established and culturally specific academic institution to countries where it had not been recognized as a norm.[13] In addition to providing institutional grants to selected research centers, like Bohr's, IEB envisioned a system of one-year international

---

[12]Pauli to Bohr, 11 February 1924 (PWB, 143).

[13]"Memorandum for the Guidance of Fellows" and "Information Concerning Fellowships in Science," 1925 (RAC. IEB. 1.3.42.599). Of course, precedents of young scientists traveling to other countries for additional training and expertise had existed before, just not as formalized and established as the institution of American postdoctoral fellowships. In Europe, countries that felt their academic system was inferior in one way or another vis-à-vis their neighbors resorted to similar methods. Thus, in 1912/13, Bohr went for additional studies in Britain after receiving his doctorate in Copenhagen, with a stipend provided by the Carlsberg Foundation. Starting mid-1850s, the Russian Empire had its own established and formalized system of educational stipends provided by the Ministry of Enlightenment and other sources, to send young scholars for training in European universities. Unlike their US colleagues, Russian professors did not consider their doctoral degrees inferior but lacked the system of what we now call "postgraduate education," and sent their students abroad before—not after—they received advanced academic degrees, i.e. on pre-doctoral fellowships.

fellowships for outstanding young academics. The board's formal rules stipulated that such fellowships could be awarded to scholars with a PhD degree or an equivalent and used only for study abroad, outside of their home country. Typically averaging $100 a month or a little more, they offered the young German doctors an extremely attractive temporary alternative (financially closer to the salary of extraordinary professor) to their struggling position at home. However, due to the international boycott, most of the foreign centers of advanced learning did not welcome students from Germany or Austria. Copenhagen, as a neutral location, was one of the few that did.

"Had Germany been more stable and prosperous, the incentive for Pauli and Heisenberg to go to Copenhagen would certainly have been less, and Bohr's influence on them might not have been so great," remarks Daniel Serwer astutely (Serwer 1977, 194). But besides the scientific attractiveness of Copenhagen and the financial support from IEB, at least one more condition was necessary: German professors, restricted in their capabilities to support younger scholars, had to grant them relative freedom to go. In March 1925, an IEB official August Trowbridge visited Munich for a discussion with professors Sommerfeld, Richard Willstätter, and Kazimierz Fajans about the means to support science. He concluded that funds for instruments were sufficient, while the assistance was primarily needed in terms of fellowships:

> Sommerfeld pointed out that in postwar Germany there exists no longer the Privat-Docent, who played a large part before the war in advanced teaching; this class of teachers was recruited from the upper "bourgeoisie"; now, that class no longer has any money and the result is that there are practically no more Privat-Docents. Professor Sommerfeld thinks that fellowships such as the IEB awards will help to keep some of the younger men connected with the universities and prevent them from going into industrial laboratories, and save them for research, just as Privat-Docent did in many cases before the war... One of the three... stated that, if there only could be a system of national fellowships in Germany... more good men could be saved for the pure science... T[rowbridge] pointed out that the I.E.B. would be quite unwilling to modify its fellowship plan so as to permit Germans to study in Germany... That was something which must be done from the country itself. All three agreed to this in principle, but sadly, as it was evident that they had hopes that the I.E.B. would make exceptions.[14]

The German academic system, however resourceful in other ways, did not change its internal rules to establish an equivalent of national postdoctoral fellowships, and professors continued to think mostly in terms of assistantships. As for the international fellowships offered by the Rockefeller philanthropy, Sommerfeld sent his students to countries that were available: Otto Laporte went to Washington, Karl Bechert to Madrid, Walter Heitler to Zurich, and Hans Bethe to Rome. Bohr started receiving IEB fellows from Germany in 1924, the majority of whom arrived by way of Göttingen, whose professors strove to establish a special relationship with Copenhagen. Nominations for fellowships typically had to be signed by one professor on the receiving end, and another one on the providing end, i.e., usually by Bohr and

---

[14]A. Trowbridge. Diaries. Log. 1, entry of March 24, 1925 (RAC. GEB. 12, on 44, 50). At a similar meeting in Göttingen on 13 March 1924, physics professors also requested no instruments, but more money for assistants. It was agreed that IEB would fund fellowships for foreign students to come to Göttingen. Born, Franck, and Pohl to IEB, 18 March 1924 (RAC. IEB. 1–2. B34. 484).

Born. Heisenberg was the first such student, beginning his IEB term in Copenhagen in September 1924; he was followed by a string of further assistants and *Privatdozenten*.

5.3 Werner Heisenberg's *Wanderjahre*

Among historians of quantum physics, Daniel Serwer was probably the first to pay serious attention to the fact that until Pauli and Heisenberg became professors, their social status did not allow them to be fully independent as researchers: "an account of their productivity in this period would take into account their intense drive for social and professional recognition; jobs were scarce and conditions unsettled" (Serwer 1977, 198). As students of and assistants to senior professors, Pauli and Heisenberg had to tune their work to the interests of their patrons. "[A]s long as I am here in Gött[ingen] I must do what Born wishes, just as in Munich I had to do what S[ommerfeld] wished," Heisenberg reassured his parents in one of his letters home (Heisenberg 2003, 56). Professors typically suggested to their "scientific helpers" (*wissenschaftliche Hilfsarbeiter*) topics for research and calculations and also, by the accepted social norm, controlled all publications coming from the institutes they directed. Prior to sending a paper to a journal, even such qualified researchers as Pauli and Heisenberg required the manuscript to be approved by the professor they worked for at the moment.

Heisenberg's personal encounters with Bohr began during the latter's lectures in Göttingen in June 1922, but his written work, letters, and publications until 1924 reveal little, if anything, of Bohr's agendas and influence. During that period Heisenberg worked for Sommerfeld and later for Born, continuing his research on complex spectra along the lines of his 1922 core model, which was in open disagreement with Bohr's views. While acknowledging the receipt of a copy of Bohr's book, Heisenberg praised Bohr's polite way of discussing the existing "disagreements about physics" (*physikalische Meinungsunterschiede*), at the very same time reporting to Landé:

> I myself am now firmly convinced, just as Professor Sommerfeld, that, in contrast to Bohr's view, the half quantum numbers are correct… Sommerfeld writes me from America that an American mathematician van Vleck calculated Bohr's model exactly and found the experimentally incorrect value of 22 V of the ionization voltage. Bohr's model must thus be wrong.[15]

Meanwhile, as he congratulated Bohr on the Nobel award, the Göttingen mathematician Richard Courant suggested the possibility that Heisenberg might come to Copenhagen to work under Bohr: "While Sommerfeld is in America young Heisenberg is here, he is a truly outstanding young man in every respect, and personally

---

[15] Heisenberg to Bohr, 14 November 1922, "Ich selbst bin jetzt auch ebenso wie Herr Prof. Sommerfeld, fest überzeugt davon, dass die halben Quantenzahlen, im Gegensatz zu Bohrs Ansicht, richtig sind… Sommerfeld schreibt mir aus Amerika, dass ein amerikanischer Mathematiker van Vleck das Bohrsche Modell gerechnet u[nd] exact den experimentell *falschen* Wert von 22 V für die Ionisierungsspannung gefunden hat. Das Bohrsche Modell muss also falsch sein." Heisenberg to Landé, 13 November 1922, English translation in (Serwer 1977, 210–11).

extremely pleasant... I am writing this because I promised you that I would bring to your attention people I believe you might sooner or later consider taking on as assistants." Bohr sent an encouraging message through Franck: "I am very interested in Heisenberg, of whom I have the best impression in all respects. Please thank him for his letter and tell him that I should always be very glad to hear from him and about his work." In June 1923 Pauli mentioned to Sommerfeld that Bohr was considering inviting Heisenberg to Copenhagen ("the pecuniary means for this are available, in any case").[16]

The rapprochement between Bohr and Heisenberg began toward the end of 1923, after Bohr's return from America, through the ritual of requesting professorial approval for a publication. Pauli, who must have discussed his Copenhagen experiences with Heisenberg, served as mediator. Heisenberg in Göttingen had completed a revised version of his core model theory, which included Pauli's skepticism about visualizing electron orbits, retained half-integral quantum numbers, but no longer explicitly contradicted Bohr's building-up principle. Sending a copy of the manuscript to Pauli, Heisenberg expressed his wish to secure a "papal blessing" (*päpstlichen Segen*) before publication. Through Kramers, Pauli checked the matter with Bohr, and finally, Heisenberg sent Bohr a detailed letter (but not the entire manuscript), which explained his main ideas at length and asked for advice as to "whether you believe that it is worth continuing to seek along this path."[17]

Bohr apologized for having had little time to think carefully about the issues involved and invited Heisenberg for a short visit, also inquiring about his long-term plans. "With regard to my plans for the future, the only thing certain is that I would like to do my *Habilitation* here with Professor Born in the foreseeable time. I would be very pleased if in the meantime I could spend a semester studying with you. Professor Born is also in agreement with this," assured Heisenberg. He was glad to come for a short visit to Copenhagen in March 1924, during which he discussed his manuscript with Bohr and made arrangements for the subsequent longer stay supported by an IEB fellowship.[18]

Bohr finally granted his approval to Heisenberg's paper in June 1924, which marked the beginning of his direct influence on Heisenberg's research. When in September that year Heisenberg arrived in Copenhagen, he had to adapt his research approaches to the new professor's agenda and preferences, just like before. The main topic in Copenhagen during that year was the BKS theory of radiation and the concept of virtual oscillators, both of which were new themes for Heisenberg.

---

[16]Courant to Bohr, 8 December 1922; Bohr to Franck, 29 December 1922; Pauli to Sommerfeld, 6 June 1923 (PWB, 95).

[17]Heisenberg to Pauli, 9 October 1923; 7 December 1923 (PWB 125–7, 132); Pauli to Kramers, 19 December 1923; "ob Sie glauben dass es sich lohnt, auf diesem Wege weiter zu suchen." Heisenberg to Bohr, 22 December 1923.

[18]Bohr to Heisenberg, 31 January 1924; "Was meine Zukunftstpläne betrifft, so steht nur soviel davon fest, dass ich mich in absehbarer Zeit hier bei Herrn Prof. Born habilitieren möchte. Wenn ich dazwischen einmal ein Semester bei Ihnen studieren könnte, so würde mich das sehr freuen. Hr. Prof. Born ist damit auch einverstanden." Heisenberg to Bohr, 3 February 1924; Bohr to Heisenberg, 18 February 1924.

Working along the lines of the "true faith," Heisenberg helped Bohr write a note on the polarization of fluorescent light, composed his own, more technical paper on the same topic, and coauthored a paper with Kramers on the theory of dispersion (Bohr 1924, Heisenberg 1925a, Kramers and Heisenberg 1925). He also often stepped in for Kramers as Bohr's "helper," the person to whom Bohr dictated letters, notes, and papers.[19]

Just before returning to Göttingen in April 1925, Heisenberg finished a paper on his earlier subject, complex spectra in multielectron atoms. This time the paper combined parts of his earlier core model approach with Pauli's exclusion principle. Bohr's influence can be seen mainly in its general discussion section and interpretation of the results, reflected in Heisenberg's reference to "non-mechanical force." Attempted as a compromise, the work did not become a breakthrough (Heisenberg 1925b). Heisenberg sent Bohr proof sheets of the paper for final approval, but Bohr often spent enormous time and effort on copyediting to get all the interpretative nuances right. Having not heard back for a month, Heisenberg grew impatient. He was also hit by an allergic attack of hay fever and had to flee for temporary refuge on the remote island of Heligoland. Before his departure, Heisenberg returned corrected proofs to the journal for print and had to apologize profusely to Bohr for not having waited longer for his master's blessing.[20] Bohr took the matter of approving papers for publication very seriously and did not easily tolerate violations of the ritual. Just like German professors, he accepted responsibility for all research performed at his institute, as well as for work that originated in it but was completed elsewhere, after the visitor's departure.

It is revealing to analyze from this perspective Heisenberg's next paper, the seminal one that launched quantum mechanics (1925c). At the time, Heisenberg was holding practically a double appointment: he returned to Göttingen in April 1925 for one semester to fulfill his duties as *Privatdozent* and substitute Born at lectures, but he continued to receive his IEB money through Copenhagen and was planning to go back there for another month in the fall, to officially complete his fellowship. The path of a peripatetic postdoctoral fellow, made possible by German inflation and Rockefeller money, submitted Heisenberg to a variety of intellectual encounters and influences: Sommerfeld's thorough training in the mathematical methods of quantum theory, Born's proposal of a fully discontinuous quantum mechanics, Bohr and Kramers's language of virtual oscillators, Pauli's rejection of visual models of electron orbits, and Ralph Kronig's calculations of the intensities of spectral lines. It is quite symbolic, however, that in order to accomplish his most daring intellectual breakthrough, Heisenberg had to escape from the authority of his academic superiors into the temporary loneliness and freedom on a small island in the North Sea. The resulting paper "On the quantum-theoretical reinterpretation of the kinematical and

---

[19]On Heisenberg as helper, see Bohr to Kramers, 23 September 1924 (BCW 5: 51–54).

[20]The paper (Heisenberg 1925b) is analyzed in Serwer (1977, 237–48); on returned proofs and Heisenberg's bad conscience, see: Bohr to Heisenberg, 10 May 1925; Heisenberg to Bohr, 8 June 1925.

Fig. 5.1 Heisenberg, 1927 (NBA)

mechanical relations" deviated from the approaches then favored at either Copenhagen or Göttingen; still, Heisenberg needed a professor to approve his manuscript for publication (Figs. 5.1 and 5.2).[21]

Born was a natural choice, but Heisenberg could have also chosen to send the manuscript to Bohr, as he had done earlier. Pauli sounded unusually enthusiastic and encouraging—atypical for this reputed critic and destroyer of the great ideas of others. Heisenberg himself was far less confident that he had accomplished something truly serious, and understandably even less confident of getting Bohr's approval for the paper. He briefly returned to Göttingen prior to leaving for another trip to give colloquium talks in the Netherlands and Britain, and a subsequent vacation. Before his departure, Heisenberg left the manuscript with Born to decide its fate. Born pondered the paper for a while, then suddenly saw how he could make an important new step in developing Heisenberg's proposal further and forwarded it for publication. It is not clear whether Bohr was informed about these developments at all; if so, then only through Kramers, who briefly visited Göttingen in the interim. As for Heisenberg, he sent Bohr neither the manuscript nor proofs nor a request to approve the publication, but only the following casual and disingenuously humble hint in a letter:

> In the entire past month I have not thought about physics at all, and I do not know if I still understand anything about it. Before, as Kramers may have told you, I had concocted a paper

---

[21] For the content and analysis of Heisenberg's paper, see Mackinnon (1977, 137–88); Darrigol (1992, 260–84).

Fig. 5.2 Bohr and Pauli, 1929 (NBA)

about quantum mechanics, about which I would be glad to hear your opinion. It is expected to appear in the next issue of the *Zeitschrift [für Physik]*.[22]

5.4 Quantum Mechanics Community

The term "quantum mechanics" (*Quantenmechanik*) had been coined by Born a year earlier as a hint toward a future "truly discontinuous" theory of atomic processes (Born 1924). Its use in Heisenberg's reference to his own paper reveals that he perceived his new approach as closer to Göttingen rather than Copenhagen, at least rhetorically. Born, too, apparently sensed the affinity and quickly moved to strengthen it further by reinterpreting Heisenberg's new rules of multiplication for quantum variables as mathematical operations with discrete matrices. Even before Heisenberg's original paper appeared in print, Born and his assistant Jordan were engaged in writing their own contribution, which they finished in September 1925. Immediately thereafter, they began collaborating with Heisenberg on a joint paper by the three authors, the so-called *Dreimännerarbeit*, which further developed the approach

---

[22]"Freilich hab ich im ganzen letzten Monat garnicht an Physik gedacht und weiss nicht, ob ich noch etwas davon verstehe. Vorher hatte ich, wie Ihnen Kramers vielleicht erzält hat, eine Arbeit über Quantenmechanik verbrochen, bei der ich gerne Ihre Ansicht hören möchte. Sie wird wohl in nächsten Heft der Zeitschrift erscheinen." Heisenberg to Bohr, 31 August 1925.

into what would become known as "matrix mechanics," the first version of quantum mechanics and Göttingen's claim for its own, truly original and independent program in quantum theory (Born and Jordan 1925; Born et al. 1926). Pauli, who was kept informed of the latest developments, proudly reported the breakthrough to Bohr as the "new Göttingen theory."[23]

Initially, Bohr had little connection to the new quantum mechanics. His favored approaches seemed to have run out of steam, while other ideas, such as light quanta and Pauli's *Zweideutigkeit* were, in Bohr's view hardly compatible with the general requirements of his correspondence principle (Bohr 1925). If not necessarily the entire old quantum theory, then at least the Copenhagen approach to it, hit an impasse. Two major initiatives appeared independently in the second half of 1925 to break the stalemate: from Göttingen, Heisenberg's quantum mechanics and from Leiden, the idea of electron spin suggested by Ehrenfest's students George Uhlenbeck and Samuel Goudsmit. Both proposals seemed to seriously contradict one another, as well as Copenhagen's, at least in spirit, but Bohr, after some hesitation, decided to endorse both of them, perhaps because at the time he lacked an alternative hope to resolve the mounting difficulties (Bohr 1925, 1926, in BCW 5, 273–280, 289). From that moment on, his primary role in the development of quantum theory would shift to that of a nurturer of ideas proposed by others, typically younger physicists, rather than by himself. As the first step in this direction, Bohr hired Heisenberg as the successor to Kramers.

Kramers was overdue to receive his own professorship, but in 1924, Bohr was still reluctant to let him go back to the Netherlands and felt content that Coster, rather than Kramers, received a position in Groningen. In the fall of 1925, Kramers accepted a call to Utrecht.[24] Heisenberg was then in Göttingen, completing the *Dreimänner-arbeit* with Jordan, while their professor and third coauthor, Born, departed for a lecture tour in America. Bohr offered the now vacant post of university lecturer to Heisenberg, simultaneously informing Franck, the remaining supervisor, about the offer. Franck was definitely upset. He and Born did not mind sending Heisenberg to Copenhagen as a postdoctoral fellow but assumed that the latter would return to Göttingen and fulfill his teaching duties as a newly minted *Privatdozent*. The prospect of losing the young star immediately after he had produced such a fundamental breakthrough was extremely disappointing to the Göttingen professors. But financially, and as a regular salaried job, the lectureship in Copenhagen certainly looked more attractive than the German *Privatdozentur*. From overseas, it was hard for Born to organize a competitive university counteroffer for Heisenberg on such short notice. After Bohr's visit to Göttingen in December 1925, Heisenberg agreed to start working in Copenhagen as of May 1. Born and Franck apparently could not object, but from then on their relationship with Bohr became somewhat strained and

---

[23] Pauli to Bohr, 17 November 1925 (PWB, 257). Pauli had strongly encouraged Heisenberg's new approach already in gestation, and he quickly moved to write his own contribution, the calculation of the hydrogen atom according to matrix mechanics.

[24] Zernike to Bohr, 25 September 1924; 29 September 1924; Bohr to Zernike, 27 September 1924; 2 October 1924; Utrecht University faculty to Bohr, 3 June 1925; Bohr to Ehrenfest, 14 October 1925.

they were more mindful of their own interests vis-à-vis Copenhagen.[25] To provide his students with IEB fellowships, Born continued sending them to Copenhagen, but was no longer satisfied with the status of a junior partner and strove to develop Göttingen as an independent and equally important center of quantum theory, with its own international postdoctoral fellows and with the new quantum mechanics as its signature brand. But since the originator of the new theory, Heisenberg, now worked in Copenhagen, Bohr's institute, too, could lay claim to the developing field.

In spring 1926, quantum mechanics entered its year of *Sturm und Drang*, with the two main approaches—Heisenberg's matrix mechanics and Schrödinger's wave mechanics—publishing more than a dozen new papers every month (Kozhevnikov and Novik 1989). Attracted by the lure of Rockefeller fellowships and often citing their interest in the new theory and the presence of Heisenberg, more students from Germany, USA, and other countries wanted to come to Copenhagen. The number of visitors who worked at the Bohr institute jumped from approximately ten in the year just prior to 1925 to more than 20 annually between 1925 and 1932, although not all of them worked on quantum mechanics or stayed long enough to produce a publication (Marner 1997). In absolute numbers, the increase may still look relatively modest, at least by today's standards, but it included the very best students in the field and, no less importantly, qualitatively changed the institute's *modus operandi*. The visitors, even as foreigners, were no longer as isolated and insecure as earlier ones tended to be, now socialized in a company of peers with analogous status. Professors, also in Göttingen and Munich, which soon also started attracting coteries of visitors, could not control these young scholars too tightly. The recognized achievements of Pauli and Heisenberg during their attempts to become more independent set up new precedents, and professors were no longer assigning topics and prescribing specific research directions as authoritatively as before. The peripatetic nature of postdoctoral life and frequent transitions between places provided an additional degree of freedom from any particular approach preferred in one single locality. Instead of a small group guided by the director's research program, Bohr's institute turned into a clearing house for emerging ideas. One can say that the postdoctoral community partially spilled out of control and was now able to run its own show, with research activity driven by collective efforts and with the trappings of youth culture.

In those two crucial years, between his endorsement of Heisenberg's initial proposal in the fall of 1925 and his work on the complementarity principle in the fall of 1927, Bohr practically did not publish. This does not mean he was absent from the story, but that his main role consisted of something else. He was almost totally preoccupied with institutional and administrative matters—working with architects, overseeing the construction of a new institute building that allowed more room for the increased numbers of visitors, corresponding with foundations, raising funds, and arranging for fellowships from IEB and other sources. As director, he created the space in which the new quantum mechanics could brew, provided financial support, and allowed the postdoctoral community more intellectual freedom than in most

---

[25]Bohr to Franck, 18 November 1925; Franck to Bohr, 20 November 1925; 29 July 1926.

other places, himself presiding over rather than guiding their collective work. Individually and intellectually, he was still recovering from the damaging fallout of the BKS gamble, and mathematically, it was not easy for him to cope with the new sophisticated techniques, especially of such visitors as Dirac in 1926 and Jordan in 1927. But he still read and authorized the submission of all publications from his institute.

Heisenberg, as the second in command, took over Kramers's duties in doing calculations and giving lectures in Danish to university students. Apparently, Heisenberg was not as willing or good with dictation, and Bohr regularly asked visiting fellows to perform this job. But when a new paper was authored by one of them, Bohr would pass the manuscript to Heisenberg, as earlier to Kramers, to check the calculations and technical aspects of the manuscript, and once confirmed, engaged himself with making corrections and final revisions in the overall argument, introduction, and interpretation. On this general level of packaging, Bohr wanted to remain and still generally was in control. He also often acted as the mediator between authors and journals, deciding when a paper was ready for publication and submitting the manuscript with a cover letter to journal editors. Bohr's clearance defined the criterion for a paper to be seen as belonging to the institute, so that the eventual journal publication could refer to Copenhagen as the place where the work had been done, and the reprint would be added to the institute's official collection of its publications—several bound volumes of which are kept in its archive.[26]

Bohr's painstaking attitude to careful formulations, to the phrasing and rephrasing of nuances, his toughness—and slowness—on matters of interpretation upset some of the fellows, including Heisenberg in the spring of 1927. At that time, Heisenberg felt excited about his new paper (Heisenberg 1927) that introduced the now famous uncertainty relations and hoped to resolve the fundamental issues of the interpretation of quantum theory. He could not, however, get the paper through Bohr's censorship, who continued to mount major objections. In desperation to publish what he thought was one of his most important insights, Heisenberg apparently overstepped the rules and used Bohr's week-long absence from the institute on a skiing vacation to send the paper to the journal without his professor's permission.[27] The incident seriously rocked their relationship, but within a few months, Heisenberg would receive an offer of professorship and started preparing to leave Copenhagen anyway. German professors, especially Sommerfeld, worked hard to win Heisenberg back for Germany as quickly as possible, which became feasible with the position that became vacant in Leipzig (Cassidy 1992, 216–8, 244–6). Heisenberg was appointed a full professor and director of his own institute in early 1928, at the unprecedentedly young age of 26.

---

[26]Universitetets Institut for Teoretisk Fysik. *Afhandlinger* (NBA). For Bohr's correspondence with journal editors, see his letters to Arnold Berliner and Karl Scheel.

[27]In his fictionalized, but fully grounded in the existing historical scholarship, psychological interpretation of the relationship between Bohr and Heisenberg, Michael Frayn presented the mainstream, overly idealized picture of their harmonious, father–son type collaboration in the process of creating quantum mechanics. He did, however, perceptively point out their conflict over the submission of the 1927 paper (Frayn 1998, 55–67).

The subtlety of control in Copenhagen allowed the visiting fellows to act and behave as if there was virtually none, and Bohr participated in maintaining this appearance. Some visitors hardly needed any more advanced learning and direction, but even they profited from the intellectual challenges and informal discussions with other fellows. By the end of 1926, Jordan in Göttingen and Dirac in Copenhagen developed schemes that allowed transformations and translations between the wave and matrix mechanics as parts of the general, more comprehensive mathematical formalism. This permitted Dirac in his next paper of February 1927 to declare quantum mechanics essentially complete as a non-relativistic theory and announce the next major goal, a fusion between quantum and relativity theories: "The new quantum theory… has by now been developed sufficiently to form a fairly complete theory of dynamics… On the other hand, hardly anything has been done up to the present on quantum electrodynamics" (Dirac 1927b, 243). This new and more complicated task would take a much longer and also less triumphant path than quantum mechanics. For some others, there still remained another fundamental challenge: that of furnishing a commonly acceptable philosophical understanding of what had been accomplished in those two years. On this philosophical aspect, too, achieving a consensus would prove more difficult than on matters of mathematical formalism, and ultimately elusive. The struggle over the philosophical meaning of quantum mechanics is considered in the following chapter (Fig. 5.3).

Fig. 5.3 Carnivalesque performance at the Copenhagen conference in spring 1930. The toy cannon and trumpet were used to make repeated loud noise, in celebration of yet another string of failures and fundamental difficulties, this time of the relativistic generalizations of quantum mechanics. This "second crisis" of quantum theory in 1930–1932 raised anticipations of yet another, even more radical conceptual revolution, which, however, eventually failed to materialize. To the disappointment of many in the audience, they would not live again through a similar catharsis as during the invention of quantum mechanics

Chapter 6
Philosophical Wrangling

I look on most general reasoning in science as [an] opportunistic (success- or unsuccessful) relationship between conceptions more or less defined by other conception[s] and helping us to overlook [danicism for 'survey'] things. – Bohr (1919)[1]

6.1 The Problem with Quantum Philosophy

In one of his last recorded statements before he died in 1962, Bohr confessed to Thomas Kuhn that he had hardly any hope of achieving an understanding between quantum physicists and philosophers. He expressed the complaint in, for Bohr, unusually strong and categorical terms: "I think it would be reasonable to say that no man who is called a philosopher really understands what one means by the complementarity description."[2] As if they were aware of this charge, philosophers retaliated some 30 years later in a volume devoted to the assessment of Bohr's contribution to philosophy. In equally strong words, Don Howard expressed doubts "whether or not Bohr's philosophy of physics can be given a coherent interpretation." As Howard summarized the problem, "There was a time, not so very long ago, when Niels Bohr's influence and stature as a philosopher of physics rivalled his standing as a physicist. But now there are signs of a growing despair—much in evidence during the 1985 Bohr centennial—about our ever being able to make good sense out of his philosophical views." The *noblesse oblige* of the professional philosopher, however, did not permit Howard to give up:

> I think that the despair is premature… What is needed at the present juncture is really quite simple. We need to return to Bohr's own words, filtered through no preconceived philosophical dogmas. We need to apply the critical tool of the historian in order to establish what those words were and how they changed over time. We need to assume, at least provisionally, that Bohr's words make sense. And we need to apply the synthetic tools of the philosopher to reconstruct from Bohr's words a coherent philosophy of physics (Howard 1994, 201).

---

[1] Bohr to Darwin, draft of a presumably unsent letter, around July 1919.
[2] Bohr, interview by Kuhn, 17 November 1962 (AHQP).

© The Author(s), under exclusive license to Springer Nature Switzerland AG 2020
A. Kojevnikov, *The Copenhagen Network*, SpringerBriefs in History of Science and Technology, https://doi.org/10.1007/978-3-030-59188-5_6

In this chapter, I take up the first, historical part of Howard's advice and follow the twists and turns of quantum philosophy during the years 1925–1927, from the refutation of the BKS theory and Heisenberg's first paper on matrix mechanics to the Solvay conference of 1927 and the first open disputes between Einstein and Bohr. Simultaneous with the invention of quantum mechanics itself, a half-dozen physicists were developing rival philosophical interpretations of the not yet completed theory. Ordered by age, this select group included Einstein, Bohr, Born, Schrödinger, Pauli, Heisenberg, and Jordan. Altogether, they expressed quite a variety of conflicting philosophical views, which can be grouped around four main issues of controversy: *Anschaulichkeit-Unanschaulichkeit* (roughly translated as visualizability–unvisualizability), continuity–discontinuity, the wave–particle dilemma, and causality–acausality.

For a historian analyzing these views, the main difficulty lies not in the paucity of sources, but on the contrary, in their intimidating overabundance and contradictory nature. That most of the above-mentioned participants and also others on their behalf continued the dispute in some form for many years after 1927 further complicates the situation. They kept commenting, explaining, and restating their positions, usually without acknowledging that their views continued to shift as the times and situation changed. Not only did the authoritative spokesmen of quantum mechanics disagree with each other—sometimes openly and sometimes subtly—but even the extant record of individual prolific writers contains mutually contradictory philosophical declarations which can only be understood within their short-term context.

Consider, for example, one of the most outspoken participants, Pauli, who left an extremely detailed and well-documented manuscript record that reveals, among other things, his radical zigzags on the issue of quantum acausality. In the previous chapter, we encountered Pauli's early comment in a 1923 letter to Sommerfeld, which compared the uncertainty and unpredictability of his personal career with that of a quantum particle. Pauli's anxiety at the height of the German economic crisis could make him more inclined to think of electrons in similar terms. By 1925 he arrived at an important conclusion that the notion of their classical trajectories inside atoms should be completely abandoned, which in turn inspired Heisenberg onto the path leading to the new quantum mechanics.[3] No matter how tempting, however, one cannot assume a direct transition from Pauli's indeterministic remark of 1923 to his advocacy of probabilistic quantum mechanics three years later. The problem is that in the meantime he also made contradictory pronouncements in no uncertain terms: "I definitely believe that *the probability concept should not be allowed in the fundamental laws of a satisfying physical theory. I am prepared to pay any price for the fulfillment of this desire, but unfortunately I still do not know the price for which it is to be had.*" The above declaration of faith sprang not from the pen of Einstein in his "God does not throw dice" mood, but from Pauli writing to Bohr in November

---

[3] For a detailed account of the development of Pauli's philosophical ideas, see Hendry (1984).

1925.[4] It contradicts much of what Pauli is otherwise known for, but at the time was made as seriously and sincerely as, later on, he would express his probabilistic convictions.

We can still understand this quote in the context of its precise timing and reference. Pauli was describing to Bohr the promise of Heisenberg's new matrix mechanics while at the same time obviously taking critical aim at, without mentioning explicitly, Bohr's failed 1924 attempt to introduce the acausal principle into the foundations of quantum theory. Knowing that experimentalists had decisively refuted the BKS proposal and feeling vindicated in his devotion to the strict validity of the conservation laws, Pauli was also ready, as the above quote shows, to reject the fundamentality of the probabilistic approach altogether. His views would change once again in less than a year. In the summer of 1926, Pauli introduced the probabilistic understanding of Schrödinger's wave function and remained forever after a proponent of the statistical interpretation of quantum mechanics. Pauli's flip-flops on the issue of causality, however, convey an important lesson, namely, that philosophical pronouncements of quantum physicists, no matter how strongly expressed, should not be taken as general and long-term commitments, but as context-dependent and flexible. As a matter of fact, such drastic shifts on fundamental issues and principles were not characteristic of Pauli alone, and not only with regard to the question of causality. Rather, they can be regarded as a distinctive feature of the early quantum philosophy in general.

In order to reduce unavoidable confusion and make sense of changing allegiances, the following analysis imposes two strong chronological restrictions on the use of sources. First, it generally avoids using post-1927 texts in which physicists explained and reinterpreted their earlier views, such as the famous recollections by Bohr, Heisenberg and Born. These later accounts were developed in the context of continuing disagreements over the foundations of quantum mechanics, and they tend to add more contradictions than clarifications if used as sources of historical information about the earlier period. Second, even within the period of 1925–27, I impose a finer time scale. The theory developed so quickly that its basic principles underwent fundamental changes approximately every six months. Statements concerning its interpretation also changed at a corresponding rate and it makes little sense to use, say, Heisenberg's pronouncements of spring 1927 for the purpose of understanding what he thought and meant in the fall of 1925, or, conversely, to rely on his initial programmatic statements of 1925 as valid for the resulting mature quantum mechanics. It is possible, however, to describe the state of quantum philosophy at a given stage of characteristic six-month lengths by using only those historical sources which come from that very time period and find, on the one hand, a sufficient number of such sources, and on the other, a significant reduction in contradictions among them.

---

[4]"Auch glaube ich bestimmt, daß *in den Grundgesetzen einer befriedigenden physikalischen Theorie der Begriff "Wahrscheinlichkeit" nicht vorkommen darf*. Ich bin bereit, für die Erfüllung dieses Wunsches einen beliebig hohen Preis zu zahlen, aber leider kenne ich den Preis noch nicht, für den sie zu haben ist." Pauli to Bohr, 17 November 1925 (PWB, 260, emphasis added).

The present reconstruction depends heavily on the wealth of the existing historiography of quantum physics. Paul Forman in several papers, including the classic "Weimar Culture, Causality, and Quantum Theory," demonstrated how the ideologically laden concepts of *Anschaulichkeit*, acausality and *Individualität* entered physicists' discourse even prior to 1925 and were subsequently ascribed to quantum mechanics (Forman 1971, 1984). John L. Heilbron described the post-1927 spread of the Copenhagen philosophy with its characteristic "combination of imperialism and resignation" (Heilbron 1985). I shall concentrate on the intermediate period in the hope of establishing a bridge between these two works. John Hendry in his book on the Bohr-Pauli dialogue presented the philosophy of quantum mechanics in the making from a more or less Copenhagen perspective. In contrast, Mara Beller developed a critique of the historical myth and of the Copenhagen orthodoxy (Hendry 1984; Beller 1999). For my analysis, I use many of their insightful observations and interpretative ideas, but also disagree on some points. The reasons are generally twofold: restrictions on the use of sources explained above and my neutral stance on philosophical issues. While admiring physicists' earlier and later interpretations of quantum mechanics as exciting intellectual achievements, I do not feel committed, at least for the purpose of this study, to any particular interpretation.

One of the main conclusions may still ultimately disappoint philosophers, namely that having fulfilled the first, historical part of Howard's advice and made situational sense of the physicists' philosophical posturing, it would become clear that the last part of his proposal—i.e., to synthesize from them a "coherent philosophy of physics"—is unrealizable. Physicists' shifting views on philosophical issues can be understood in their own local times and contexts, but taken together as a set, they constitute a self-contradictory body of propositions that allows for a variety of irreconcilable interpretations. Overall, the philosophical discourse of quantum physicists was opportunistic in the sense of Bohr's quote in the epigraph to this chapter. Physicists made philosophical statements as if announcing strongly held principles, but they also kept changing them rather easily, sometimes to almost the opposite within a single year. They also used those statements as rhetorical resources in their intradisciplinary rivalry, in some cases overstating the existing differences, or downplaying and hiding them away, due to tactical reasons and personal relationships. Beller came to a similarly sounding conclusion about the impossibility of presenting the Copenhagen interpretation consistently, but only as a "contingent composite of constantly shifting differences among its founders... [P]hilosophical pronouncements by quantum physicists are most adequately understood as local, shifting, and opportunistic" (Beller 1999, 173). Rather than placing the chief blame on one side of the controversy, I see in such inconsistencies a general pattern of behavior of the entire disciplinary community. The point is not that a particular version of quantum philosophy is unsatisfactory, but that the entire interpretational debate was something else dressed up in philosophical garb. Professional philosophers' feelings of despair came not from the deficiency of their "synthetic tools," but from the a priori assumption that some consistent and coherent doctrine was hiding behind physicists' rhetoric. For a philosopher, dropping this assumption would amount to admitting that the discourse was not philosophical in the strict sense. In the conclusion, therefore, I will have to switch the mode of analysis from the history of ideas to cultural history in order to understand what kind of activity it was, if not philosophical.

6.2 Matrix Mechanics (Fall 1925)

Familiar concepts and images of classical physics were not faring well in the atomic domain. In quieter and more positive times, scientists could have remained more tolerant of the developing contradictions, but those who shared the existential experiences of life in Europe during the second decade of the twentieth century were accustomed to seeing crises and revolutions in every venue of life, including science. Often, they were more willing than reluctant to read existing problems as signs of foundational crises. The quantum theory of the atom developed since 1913 by Bohr and Sommerfeld with co-workers promised a radical solution of one such crisis at the price of abandoning some basic and proven postulates of classical mechanics and electrodynamics. After spectacular successes in understanding and calculating atomic spectra of hydrogen, the theory also encountered problems, in particular, in attempts to generalize it to the case of multielectron atoms. Again, in some other epoch, ours for example, physicists would have been more inclined to see the glass as half-full rather than half-empty, or at least allow the adolescent theory a little more time to prove itself. In the radical 1920s, however, revolutionary proposals themselves, and not just traditional beliefs, were subject to heightened degrees of criticism. By 1925, circles of physicists around Bohr in Copenhagen and Born in Göttingen came to the conclusion that the quantum theory of the atom, too, no matter how young and radical, had entered a state of foundational crisis.

To find another revolutionary solution, they were prepared for further sacrifices in the most basic principles of physics. "Most basic" to them meant philosophical, and being "philosophically minded" constituted praise within this circle. It was not quite obvious, however, what exactly had to be sacrificed. The list of possible and tried victims included, but was not limited to: (1) ideas of space and time, (2) energy conservation, (3) causal description, (4) the concept of electromagnetic field, and (5) continuity of kinematics.[5] After a number of unsuccessful attempts, they found much promise in a paper by Heisenberg (1925c) and collaborated on the theory, which became known as matrix mechanics. It existed in its original form until the beginning of 1926 with its own characteristic set of philosophical preferences.

Unanschaulichkeit. The first and most distinctive on the list, as demonstrated by Beller (1983), suggested abandoning the usual ideas about space and time. Our common visual intuitions, one could argue, relied on human experiences in the macroscopic world with objects roughly the size of our own, but did not remain valid within the microscopic domain. Trying to make sense of atomic phenomena with the help of such inadequate intuitive visual (*anschauliche*) representations could

---

[5]Darwin to Bohr, 20 July 1919. See also Hendry (1984, 20, 29, 31, 33, 36, 37, 55, 64) for relevant quotations from various authors.

be the chief source of contradictions encountered within the quantum theory of the atom. Different formulations of this idea were provided by Bohr (complete space-time representation of atomic processes is impossible), Born (geometry fails within the atom), Heisenberg (positions and trajectories of the electron in the atom do not exist) and Pauli (abandonment of the mechanical, spatial–temporal representation of the stationary state). To build a new theory from the ground up, it had "first to throw away visual representations of the atom," the *Anschaulichkeit*.[6] Not necessarily rejoicing about this feature, Heisenberg, Pauli, Born, Jordan, and Dirac accepted *Unanschaulichkeit* as the basic and necessary premise of the new theory.

Discontinuity. In matrix mechanics, the atomic world was *unanschaulich* in large part because of its fundamental discontinuity. "In processes at microscopic dimensions of space and time, a discontinuous element plays the dominant role," which could not be adequately expressed and represented with the usual, continuous space-time conceptions (Heisenberg 1926, 704). Matrix mechanics inherited not only the energy states of Bohr's early atomic theory, but also Born's 1924 program of a "truly discontinuous theory" which proposed to consistently replace all continuous physical concepts with discrete sets. In matrix mechanics, the transition from classical to quantum theory was achieved accordingly by substituting continuous variables with discrete matrices.

The following two philosophical issues did not play such a major role at the matrix mechanics stage as they had and would in some earlier and later versions of quantum theory. Their very absence is significant, nevertheless.

No waves, no particles. Since matrix mechanics and wave mechanics competed with each other, some commentators assumed that matrix mechanics favored corpuscular ontology over waves. Beller rightly criticized this view, but her assertion that matrix mechanics "was thoroughly permeated by wave-theoretical concepts" (Beller 1983, 477) is equally untenable (she supported this claim mostly with quotes from the earlier period of the BKS theory). Both waves and particles were visual representations and thus unsuited for an *unanschaulich* theory. Only outside of the atom did radiation consist of waves while electrons were corpuscles, but inside the atom, the electron and its radiation together were represented by a discontinuous and unvisualizable set of matrix elements. Neither intuitive images of waves nor of particles were useful for its description. The only exception to this attitude came in Born's American lectures of winter 1925, where he tried to combine, somewhat artificially, the *Unanschaulichkeit* of matrix mechanics with Einstein's wave theory of matter and suggested (even before wave mechanics) the existence of some undulatory process within the atom (Born 1926a).

No time, no acausality, no statistics. The idea of acausality together with the statistical conservation of energy had been tried earlier in the BKS theory of 1924. Bohr turned to that risky hypothesis in a last desperate attempt to save the wave theory

---

[6] "die Erlösung…ist um keinen geringeren Preis zu erzielen als dem des Verzichtes auf mechanische, raum-zeitliche Bilder der stationären Zustände des Wasserstoffatoms." Pauli to Bohr, 17 November 1925 (PWB, 188, 258); "sollte die neue Theorie auf die Anschaulichkeit zunächst ganz verzichten." (Heisenberg 1926, 705).

of electromagnetic radiation from the abhorrent (to him) notion of light quanta. At that juncture, Schrödinger welcomed the acausal idea, while Pauli and Einstein criticized it (the latter not yet doing so as a matter of philosophical principle, but because he had already tried it earlier without much success). Born and Franck did not feel happy about it, either, but did not want to contradict Bohr and were trying to say something polite, if vague. Heisenberg, Bohr's formal employee during that year, appeared to accept the idea on the surface, but likely not in his heart: in his papers, he used the language and approach of the BKS theory, but carefully avoided its most dangerous assumption. Refuted by Bothe and Geiger's experiment in 1925, the idea seemed to be totally discredited and did not appear in matrix mechanics at all. Pauli's comment in November 1925 to this effect, which strongly rejected the very use of probability in fundamental physical theory, has been quoted above. Heisenberg distanced his new approach from acausality by purging the very word "probability" from his matrix mechanics papers. Instead of "probability of [atomic] transitions," he consistently used "intensities of emitted radiation." The two phrases can be used interchangeably in our times, but in the context of 1925, physicists were quite sensitive to this choice of words.

Handling disagreements. The authors of matrix mechanics did not agree on some other interpretational issues. The most serious of these concerned the definition of the basic quantities of the new theory. Born defined them mathematically simply as matrix elements, thus deviating from Heisenberg's original (and not entirely satisfactory) physical definition of them as amplitudes of emitted radiation. What Heisenberg took to be the most important physical postulate of matrix mechanics does not even appear in the core of the theory in Born and Jordan's presentation. They only introduce it as an auxiliary assumption ("Heisenberg's *Annahme*") for the purpose of calculating intensities of spectral lines at the very end of their paper (Born and Jordan 1925).

This discrepancy helps to explain why Heisenberg disliked Born's matrices and was unhappy about the very name "matrix mechanics." He contrasted his "physical" approach to the "mathematical" one of the Göttingen physicists and struggled to preserve his interpretation of the theory while collaborating with Born and Jordan on the famous *Dreimännerarbeit* of November 1925. From Göttingen, he wrote to Pauli:

> I tried as hard as I could to make the theory more physical, but am only half-satisfied with the result. I am still quite unhappy about the whole thing and was so glad to hear that, with regard to mathematics and physics, you are completely on my side. Here I am in a milieu that thinks and feels exactly the opposite and I worry whether I am just too stupid to understand mathematics... I always feel irritated when the theory is called matrix mechanics and for a time seriously wanted to cross the word "matrix" completely out of the paper and replace it with, for example, "quantum-theoretical variable." (After all, "matrix" is one of the dumbest mathematical words.)[7]

---

[7]"Ich hab' mir alle Mühe gegeben, die Arbeit physikalischer zu machen, als sie war und bin so halb zufrieden damit. Aber ich bin immer noch ziemlich unglücklich über die ganze Theorie und war so froh, dass Sie mit der Ansicht über Mathematik und Physik so ganz auf meiner Seite stehen. Hier bin ich in einer Umgebung, die genau entgegengesetzt denkt und fühlt und ich weiss nicht, ob ich

Despite these private complaints, conflicts did not go public. The authors of matrix mechanics chose to collaborate on the new theory. They advanced their diverging interpretations in separate publications, but did not explicitly set them against each other and avoided discussing their disagreements in public.

6.3 Wave Mechanics (Spring 1926)

Schrödinger's first paper on wave mechanics in January 1926 cautiously emphasized formalism rather than interpretation. As another precaution, he made a friendly gesture toward matrix mechanics in mentioning that both theories had one basic feature in common: the abandonment of the notion of electron trajectories (Schrödinger 1926a). The statement was hardly sincere, because the reasons for this abandonment were different in the two theories. In wave mechanics, the electron did not have a definite position not because of *Unanschaulichkeit*, but because it was represented by a continuous wave and spread out in three-dimensional space. Once Schrödinger had become more confident of the success and power of his theory, he did not need the protective rhetoric any longer and fully engaged in the interpretation business. In March, he established a mathematical connection between the basic formulae of the two theories and proclaimed them "mathematically equivalent" (Schrödinger 1926b). This was an understatement—wave mechanics was certainly much more powerful and handy in calculations than matrix mechanics—but the implication was that the criterion for choosing between the two should be interpretation rather than formalism. At that stage, Schrödinger was confident his interpretation had to be preferred.

Complete restoration of Anschaulichkeit. Wave mechanics' main philosophical advantage appeared in the rehabilitation of *Anschaulichkeit*. Not only did the usual three-dimensional geometry remain completely valid on the microscopic scale but even the motion of the electron within the atom could be represented pictorially (the difference from classical theory being the visual image was of a vibrating string instead of a moving corpuscle). The space-time visualization of microscopic processes was declared possible again.

Continuity. In wave mechanics, discrete energy levels are obtained as solutions of a continuous wave equation. One could still, in principle, choose which one aspect to select and stress as fundamental—continuity or discontinuity—and the question turned into a heated debate in 1926. Bohr wanted to welcome wave mechanics but insisted it should be understood precisely as a description of discontinuous atomic states. Schrödinger, on the other hand, emphasized the continuity aspect alone, taking

---

nur zu dumm bin, um Mathematik zu verstehen… Ich bin immer wütend, wenn ich die Theorie nur unter dem Namen Matrizenphysik genannt höre und hatte eine Zeit lang ernstlich vor, das Wort Matrix ganz aus der Arbeit zu streichen und durch ein andres z.B. 'quantentheor[etische] Größe' zu ersetzen. (Übrigens, ist Matrix wohl eines der dümmsten Mathematischen Wörter, die es gibt)." Heisenberg to Pauli, 16 November 1925 (PWB, 255). Dirac subsequently designed a special term "q-numbers" for quantum variables.

"a departure from fundamental discontinuity" as his main philosophical slogan and programmatic goal.[8] For him, not only were discrete energy states artefacts of continuous undulatory processes, but quantum transitions themselves had to be explained as continuous changes from one vibrational mode to another, point particles had to be understood as wave packets and the very relationship between classical and quantum descriptions was to be conceived as "the *continuous* transition from micro- to macro-mechanics" (Schrödinger 1926c, emphasis added).

Wave ontology. Although de Broglie's dualistic concept of waves and particles provided initial inspiration to Schrödinger, duality did not figure prominently in wave mechanics during its heyday in the spring of 1926. Schrödinger openly and obviously preferred waves to corpuscles as ontological reality. Radiation appeared in his theory in the form of classical electromagnetic waves. Electrons were perceived as corpuscles only on the scale of lower resolution, whereas at the truly microscopic quantum scale they were wave packets of a finite size. (The difference is similar to that between the geometrical and the more fundamental wave optics.) Schrödinger hoped at the time to develop a theory in which all particle-related concepts would be replaced consistently by undulatory ones (for instance, energy would have to be replaced by frequency and the concept of quantum transitions by resonance). Such an ultimate field-like view had no need for wave-particle duality.

No statistics, no acausality. This is the only main philosophical feature that wave mechanics and matrix mechanics had in common. Their principal stakes were elsewhere, but both shared a definite dislike for statistical considerations and deliberately eluded the language of probabilities. Although in 1924 Schrödinger supported the statistical BKS theory, its defeat must have affected him, too, for, just like Heisenberg, he consistently used the term "intensities" instead of "transition probabilities." Moreover, he hoped to explain quantum transitions through a causal and continuous process: In a linear combination of vibrational modes, some coefficients would grow, while others would decrease in time, thus accounting for the gradual transition from one vibrational mode to another.

Reactions to wave mechanics. The rivalry between the two approaches has sometimes led commentators to assume the authors of matrix mechanics accepted Schrödinger's theory only reluctantly, after it found a very enthusiastic general reception among physicists. A distinction between happiness and quickness can provide a more accurate perspective. The captains of matrix mechanics were among the first to abandon their sinking vessel and to start using the new methods of wave mechanics, although in ways that often transcended the boundaries of Schrödinger's original intent (Kozhevnikov and Novik 1989). Pauli was the quickest: he learned of the new achievement from Sommerfeld, and in April 1926, simultaneously with Schrödinger, proved the mathematical equivalence of the two theories. Born was happiest: He easily and enthusiastically converted to wave ontology in his papers of summer 1926. Heisenberg was the unhappiest, but even he used wave functions in

---

[8]"eine Abkehr von den grundsätzlichen Diskontinuitäten." Schrödinger to Wien, 18 June 1926 (DM).

his June 1926 paper. Only Dirac was slow, first turning to Schrödinger's methods in August 1926.

Their reaction to the philosophy of wave mechanics was certainly much more critical, but even here some of Schrödinger's accomplishments could not be resisted. *Anschaulichkeit* had to be rehabilitated, at least partially. Much of matrix mechanics' former radical opposition to visualization of atomic processes quietly disappeared from its authors' subsequent publications in the course of 1926. Besides wave mechanics, another visual concept also contributed to this change of heart: The proposal of the spinning electron gained quick acceptance, despite the initially skeptical reception by Pauli and Heisenberg. At the end of the day, Euclidian geometry did not fail within the atom and visual pictures of microscopic processes proved, once again, their heuristic usefulness. *Unanschaulichkeit* retained some territory: quantum transitions, or mysterious jumps, avoided visualization despite Schrödinger's initial hopes. But it became increasingly hard to insist on it as a grand philosophical principle, although Heisenberg (with some assistance from Bohr) continued his desperate struggle against visualization until the spring of 1927 and his own paper on the indeterminacy principle. A better strategy was to hide the philosophical defeat by shifting the public debate to other issues of controversy.

The wave ontology appealed to at least some of the matrix people. Born, who had liked Einstein's idea of matter waves even earlier, subscribed to it enthusiastically. Bohr was also quite sympathetic, and Pauli did not particularly object. Heisenberg was as unhappy about waves as just about all other physical ideas of wave mechanics. He wanted to deprive ψ of its physical meaning as a wave and reduce it to a mere mathematical tool. Dirac also preferred particles to waves and the treatment of the wave function as an abstract mathematical symbol.

The entire group united in opposition to Schrödinger's continuity claim. Born's main contribution to wave mechanics was to show that it "permits description not only of the stationary states, but also of quantum jumps" (Born 1926b; Beller 1990, 567; 1992). Pauli wrote to Schrödinger in May 1926: "I have generally the strongest doubt in the feasibility of a consistent wholly continuous field theory of the de Broglie waves. One must probably still introduce into the description of quantum phenomena essentially discontinuous elements as well."[9] The stated goal of Heisenberg's two papers of summer 1926 was to prove the essential discontinuity of atomic phenomena, even when described by the Schrödinger function. And famously, the crucial dispute between Bohr and Schrödinger during the latter's visit to Copenhagen in September 1926 centered on their main disagreement about discontinuous quantum jumps.

---

[9]"ich habe überhaupt die stärksten Zweifel an der Durchführbarkeit einer konsequenten reinen Kontinuums-Feld-Theorie der de Broglie-Strahlung. Man muß wahrscheinlich doch auch wesentlich diskontinuerliche Elemente in die Beschreibung der Quantenphänomene einführen." Pauli to Schrödinger, 24 May 1926 (PWB, 326), English translation from Hendry (1984, 86).

6.4 Quantum Mechanics (Fall 1926)

While appropriating Schrödinger's wave mechanics, Born, Pauli, Heisenberg, Dirac and Jordan did not feel bound by his original interpretations but applied the theory quite liberally to new kinds of problems, thereby changing the meaning of its basic concepts. By generalizing the approach to treat the multielectron problem, Heisenberg and Dirac transformed ψ into a wave function in multidimensional space, which eroded its initial visual interpretation as a wave in ordinary space. By applying the method to the problem of scattering, Born, Pauli and Dirac changed ψ into a guiding field for particles and into a probability distribution, once again depriving it of its original physical meaning. By the end of 1926, Dirac and Jordan unified all these new accomplishments into a general scheme under the name of transformation theory and declared the (non-relativistic) quantum formalism completed. Their decisive synthesis brought about further shifts in philosophical positions (Dirac 1927a; Jordan 1927a; Ehlers et al. 2007).

Limited Anschaulichkeit. The common perception that Schrödinger lost his philosophical struggle overlooks the major fact that he had basically won the battle for *Anschaulichkeit.* Objections to ordinary geometry, the usual ideas of space and time, and to visual pictures with either waves or particles disappeared. Born used all these notions in his papers on scattering in wave mechanics. Pauli made a further concession and a reversal of his earlier cherished beliefs when he rehabilitated the notion of the "position of the electron within the atom," the probability of which was now determined by the wave function. Without an open admission of failure, the main programmatic claim of the initial matrix mechanics was dropped and disappeared from the discourse. However, the restoration of *Anschaulichkeit* did not become absolute: probabilistic arguments imposed restrictions on it. The theory permitted calculation only of the probabilities of the electron's position and of the still-*unanschauliche* quantum jumps.

Symmetry between continuous and discontinuous representations. In Copenhagen in September 1926, while Bohr and Schrödinger conducted their very intense, principled and stubborn disputes about continuity and discontinuity in atoms, Dirac quietly worked on a paper that would render this entire polemic obsolete. Unlike the rest of the group, Dirac did not label his ideas "philosophical," but his reformulation of the basic principles of quantum mechanics affected others' philosophical reasoning. Dirac developed a mathematical formalism in which both continuous and discontinuous quantum variables could be used in a relatively symmetrical fashion. His theory allowed transformations from one set of variables to another, thus putting them on an equal footing (Dirac 1927a). The continuity–discontinuity dilemma thus turned into a choice determined by simple mathematical convenience regarding which particular variables could work better for calculating one or another problem in atomic physics. It no longer made much sense to treat it as a matter of philosophical gravity.

Duality. Following Born's reinterpretation of the wave function as a guiding field for particles, both wave and particle visualizations of microscopic events began to be used, frequently and often interchangeably, in quantum mechanics. Some physicists preferred one over the other, but the discipline as a whole demonstrated a rather promiscuous use of both corpuscular and wave pictures (partly justified by the transformation theory, although Dirac personally always gravitated toward particles). A physicist could use one or both of these visualizations as intuitively helpful, suggestive pictures, but to take either of them literally and sincerely, would mean pushing the matter too far. Asking for a disciplined usage or clear choice between them looked increasingly pedantic and old-fashioned. We may call such widespread carelessness and libertarian use of either wave or particle language with inconsistent switches from one to the other "duality" to distinguish it from rarer occurrences of "dualism," or serious statements about the ontological reality of wave–particle chimeras (Kojevnikov 2002).

Causality and statistics. With the erosion of earlier philosophical principles, a new, statistical idea was on the rise in the fall of 1926 through the contributions of Born, Pauli, Dirac and Jordan. In the corpuscular representation, the wave function determined probabilities of the electron's states and transitions. In August 1926, on the eve of his Copenhagen visit, Schrödinger explained in a letter to Wilhelm Wien his standing on the interpretational issues. Schrödinger rejected *a limine* "Bohr's standpoint, that a space-time description is impossible," but showed somewhat more understanding for Born's emerging statistical picture:

> Today I no longer like to assume with Born that an individual process of this kind is "absolutely random," i.e. completely undetermined. I no longer believe today that this conception (which I championed so enthusiastically four years ago) accomplishes much. From an offprint of Born's last work in the *Zeitschr.f.Phys.* I know more or less how he thinks of things: the *waves* must be strictly causally determined through field laws; the wave functions, on the other hand, have only the meaning of probabilities for the *actual* motions of light or material particles. I believe that Born overlooks that—provided one could have this view worked out completely—it would depend on the taste of the observer *which* he now wishes to regard as *real*, the particle or the guiding field. There is certainly no criterion for reality if one does not want to say: the *real* is only the complex of sense impressions, all the rest are only pictures.[10]

Schrödinger was thus prepared for a compromise, on positivistic terms, between the wave and the corpuscular, the causal and the statistical, interpretations of the

---

[10]"Ich möchte aber heute nicht mehr gern mit Born annehmen, dass solch ein einzelnes Ereignis 'absolut zufällig' d.h. vollkommen undeterminiert ist. Ich glaube heute nicht mehr, dass man mit dieser Auffassung (für die ich vor vier Jahren sehr lebhaft eingetreten bin) viel gewinnt. Aus einem Probeabzug von Borns letzter Arbeit in der Zeitschr. f. Phys. weiss ich ungefähr, wie er sich die Sache denkt: die *Wellen* sollen streng kausal durch Feldgesetze determiniert sein, die Wellenfunktionen sollen aber ihrerseits nur die Bedeutung von Wahrscheinlichkeiten haben für die *wirkliche* Bewegung der Licht- oder Materiepartikeln. Ich glaube Born übersieht dabei, dass—angenommen dieses Bild läge vollkommen durchgearbeitet vor—es immer noch dem Geschmack des Beschauers anheimgestellt bleiben würde, *was* er nun als *real* ansehen will, die Partikel oder das Führungsfeld. Ein philosophisches Kriterium der Realität gibt es ja nicht, wenn man nicht sagen will: *real* ist einzig der sinnenfällige Komplex, alles andere sind nichts als Bilder. Bohrs Standpunkt, eine räumlich-zeitliche Beschreibung sei unmöglich, lehne ich a limine ab." Schrödinger to Wien, 25 August 1926 (DM). I am thankful to Cathryn Carson for a copy of the original text. The English translation is partially borrowed from Moore (1989, 225–26).

theory. If one were inclined to accept waves as the ultimate reality, the fundamental laws of the theory would be causal. If the fundamentality of particles was assumed, their laws of motion would be probabilistic. Schrödinger preferred the former option but was willing to put up with those who gravitated toward the latter. Born's position at first, as expressed in his July 1926 paper on probabilistic scattering, seemed compatible. He personally liked the corpuscular and acausal picture rather than the one with waves and causality, but regarded this still as a matter of philosophical taste, not principle: "I myself am inclined to renounce determinism in the world of atoms. But that is a philosophical question for which physical arguments alone are not decisive" (Born 1926b; Beller 1990; Greenspan 2005, 139).

6.5 Philosophies of Compromise (1927)

In the fall of 1926, three centers could compete for leadership in the new quantum mechanics. In Copenhagen, Bohr was still silent in public, but he hired Heisenberg, who kept on publishing important papers, attracting new visitors to the institute, and on the philosophical front continued to defend the remains of the matrix mechanics agenda (*Unanschaulichkeit* and discontinuity). Schrödinger promoted wave mechanics and the ideas of wave ontology and continuity in Zurich. In Göttingen, where the whole thing started, Born was determined to maintain momentum despite the damaging loss of Heisenberg and, together with Jordan, was developing the probabilistic version of quantum mechanics. The following year, new philosophies appeared which drew upon the earlier ideas in more complex and mixed ways.

Born's move toward acausality. Approximately once a year Bohr invited a distinguished visitor to his Copenhagen institute. Extending such an invitation to Schrödinger indicated Bohr's interest in an agreement, cooperation and a possible deal, rather than a quarrel.[11] Indeed, during their week-long non-stop Copenhagen discussions, Bohr did not push hard on *Unanschaulichkeit* and was very sympathetic to the wave mechanics in general and the wave ontology in particular. In return, he wanted Schrödinger to retreat on the maximalist continuity claim and accept the fundamental discreteness of atomic phenomena. A compromise along these lines would have included a fusion of the wave mechanics with discontinuous quantum states and jumps of Bohr's original theory of the atom:

> A few weeks ago we had a visit of Schrödinger, which gave rise to much discussion regarding the physical reality of the postulates of the atomic theory. I suppose you know that the wonderful results Schrödinger has arrived at has led to the suggestion, taken up with great enthusiasm from various sides, that the ideas of discontinuity which underlie the interpretation hitherto given of the phenomena might be unnecessary. This appears, however, to be a misunderstanding, as it would seem that Schrödinger's results so far can only be given a physical application when interpreted in the sense of the usual postulates. Indeed they offer a

---

[11]Bohr to Schrödinger, 11 September 1926; Schrödinger to Bohr, 21 September 1926.

most welcome supplement to the matrix mechanics in allowing to characterize the stationary states separately.[12]

Schrödinger, however, refused to accept discontinuity as stubbornly as Bohr insisted upon it. As we saw above, he preferred a compromise with Born rather than Bohr. Born, for his part, resolutely declined Schrödinger's advances. As professorial wrangling often goes, he stood behind while encouraging his *Privatdozent* Jordan to launch an open attack in print on the philosophy of wave mechanics. Schrödinger tried to smooth out the relationship and complained about this critique in a private letter to Born, which the latter ridiculed in his private circle. Early in 1927 Born and Jordan publicly proclaimed acausality as the most important philosophical lesson of quantum mechanics (Jordan 1927b; Born 1927; Beller 1990, 572–73). Relying on the new formalism of the transformation theory, they explicitly criticized Schrödinger's wave ontology. On the other hand, their philosophy had room for Copenhagen's favorite discontinuity, thus making possible a compromise with Bohr.

Schrödinger's move toward wave–particle dualism. In the fall of 1926, Schrödinger was named the second choice (after Sommerfeld, but before Born) in the search to succeed Planck in the most prestigious chair of theoretical physics at the University of Berlin. After Sommerfeld declined as anticipated, Schrödinger accepted the offer and moved to Berlin (his former position in Zurich would subsequently become Pauli's). A win on prestige, however, eventually turned into an institutional disadvantage for Schrödinger. In subsequent years, he worked in relative isolation, usually with only a couple of assistants in Berlin, while much larger and more active research communities of younger students and postdoctoral visitors grew around Göttingen and Copenhagen (and also later around Heisenberg in Leipzig). In philosophical terms, Schrödinger moved toward an open critique of the statistical interpretation after Born had rejected a possible compromise: "Personally I no longer regard this [statistical] interpretation as a finally satisfactory one, even if it proves useful in practice. To me it seems to mean a renunciation, much too fundamental in principle, of all attempt to understand the individual process." Eventually, Schrödinger would also retreat from a strong wave ontology and, together with his *Privatdozent* Fritz London, embrace wave–particle dualism. London's lectures on wave mechanics in Berlin opened with a programmatic statement on the dual (wave and particle) nature of quantum objects.[13]

Heisenberg's move to indeterminacy. Born's acausality met with mixed reactions in Copenhagen. Heisenberg welcomed statistics as an argument against Schrödinger's philosophy, but both he and Bohr preferred to view it as a part of mathematical "formalism" rather than the philosophy of quantum mechanics, which both

---

[12]Bohr to Kronig, 28 October 1926 (ETH). See analysis of these discussions in Beller (1992).

[13]Schrödinger (1927, 272), English translation in Schrödinger (2003, 135–36); Fritz London, "Quantenmechanik, insbesondere Anwendungen auf die Mehrkörperproblem u.d. Chemie," unpublished lectures at Berlin University, 1928–29, (Fritz London Papers, Duke University Archive). Mindful of political sensitivities, Rockefeller philanthropists were reluctant to support large institutional projects in the imperial capital, Berlin, but were more open to invest in academic centers elsewhere in Germany.

intended to develop on their own. The resulting Göttingen–Copenhagen alliance, if it can be called that, formed out of convenience. On the basis of the shared formalism of quantum mechanics, its major spokesmen advanced de facto diverging interpretational claims, but did not criticize each other's views in public, maintaining at least a posture of good cooperation.

Heisenberg, still Bohr's subordinate, refused to wait patiently. In his famous paper of March 1927, he argued that the statistical formalism led to a fundamental philosophical consequence: the unavoidable uncertainty in the simultaneous measurement of a particle's position and velocity (Heisenberg 1927). Although not quite so radical as the *Unanschaulichkeit* claim of the earlier matrix mechanics, it imposed a fundamental restriction on the visualizability of classical theories. In a letter to Kronig, Heisenberg summarized the combination of philosophical themes of his work as follows: "I have recently done a paper about the visualizable content of the (certainly discontinuous) quantum mechanics, which presents my (or all of us here's) view based on the now completed scheme and also answers the question: light quanta or waves. You will see it in the *Zeitschrift*!"[14]

Bohr's move toward complementarity. Bohr considered Heisenberg's uncertainty paper premature and they argued intensely over the manuscript. After Heisenberg sent it out for publication without Bohr's approval and later that year accepted a professorship in Leipzig, it took time to heal, but not completely, their somewhat strained relationship.[15] After two years of public silence, Bohr felt obliged to respond to Heisenberg. Now that both the construction of a new institute building and the formal edifice of the new quantum mechanics was complete, he started developing his own interpretation of this fundamental breakthrough. Bohr's writing proceeded, as usual, slowly and required a helper with whom he could collaborate on discussing the manuscript and dictation. Always struggling to arrive at definitive formulations and almost never fully satisfied, Bohr went through multiple revisions and proofs. With the help of Klein, he completed the manuscript by the end of 1927.

Bohr's interpretation is complex and difficult to understand, in part because it draws on everybody else's, as if trying to ensure all important contributors would find something in it they personally cherished: Bohr's own favorite discontinuity, Schrödinger's wave packets, Heisenberg's *(Un)anschaulichkeit* and indeterminacy, and Born's acausality. According to Bohr, there is a fundamental discontinuous, somewhat mystical, individuality (*Individualität*) at work in all microscopic processes and our imperfectly human means of comprehending it. When trying to make sense of atoms, one cannot help but alternate between visual space-time and causal-logical descriptions of events. Both intuitions derive from classical physics and our macroscopic experiences and are therefore not entirely suitable for describing

---

[14]"Ich selbst hab in den letzten Monaten eine Arbeit über den anschaulichen Inhalt der (natürlich diskontinuerlichen) Quantenmechanik gemacht, die meiner (oder unserer aller hier) Ansicht nach endlich ein geschlossenes Schema darstellt, das auch die Frage: Lichtqu. oder Wellen beantwortet. Aber Sie werden ja in der Zeitschr. lesen!" Heisenberg to Kronig, 8 April 1927 (ETH).

[15]"I was so unhappy last winter, how everything became estranged and how ungrateful I seemed towards you... I hope you can forgive everything that I have done wrong." Heisenberg to Bohr, 21 August 1927. See also Heisenberg (2003, 121–22).

the strange microscopic world. But experimental settings, insofar as they involve macroscopic instruments, make the use of such classical languages unavoidable. An uncontrollable disturbance of the microscopic object in the process of observation imposes limits on their applicability, however. An experimental setup designed to investigate and determine the space-time picture of microscopic phenomena makes impossible their causal representation and vice versa. In attempting to combine them too literally in quantum physics, as classical physics was able to, one becomes mired in inevitable contradictions: It is thus necessary to renounce the possibility of their simultaneous unlimited application within the quantum domain. Each representation separately is also insufficient for understanding the full range of possible experience with atoms, but every imaginable experiment can be accounted for in terms of one or the other description. Though based on conflicting sets of notions, these representations should be taken not as mutually exclusive, but complementary—only their combined, alternating use can produce the fullest possible account of the microscopic world (Bohr 1928; Murdoch 1987; Heilbron 1985, 199–200).

6.6 Discussion

Having followed the twists and turns of the philosophical discourse, we can see it as simultaneously high-principled—participants were utterly serious in making strongly worded philosophical statements; relatively undisciplined—their conceptions only "more or less defined by other conceptions"; and opportunistic—the proclaimed principles kept changing too often. It is hard to avoid the impression that physicists acted as if compelled to hurry up in declaring general philosophical conclusions, which often happened to be premature, because the theory itself was still *in statu nascendi*. It appears that advancing a philosophical interpretation was an invaluable act in itself, apart from the choice of a particular philosophy or the probable time for it to hold. Such an idiosyncratic behavioral pattern calls for a cultural interpretation.

Some background aspects of the phenomenon, at its most basic and obvious level, are not uncommon, but recognizable as typical and natural for a specific social group, the German academics, or *Gelehrte*. The culture of German-speaking academe upheld the strong ideal of a scientific genius forced to partially double as a philosopher. A truly great scientist was expected not only to make discoveries in a special field of research, but to go into it in such depth as to contribute to a general philosophical outlook, and to such conclusions that would be meaningful to all members of educated culture, transcending narrow professionalism and disciplinary boundaries. To this widely shared belief we owe the abundance of printed talks and *Habilitation* speeches addressed to general academic audiences, in which German scholars discussed broader cultural meanings of their special field of study.[16]

---

[16]This genre of writing on philosophical lessons from science provided abundant sources for the Forman thesis on Weimar culture and quantum acausality (Forman 1971). For a discussion of the

For the purposes of current discussion, we can take it for granted as a well-established and entrenched ritual, which in the case of quantum mechanics, however, produced an atypical outcome.

The very scope of the debate was already unusual—it would be hard to point out another scientific development in which the existing genre of philosophizing produced an intellectual fight of such intensity and inconsistency of positions among such a number of prominent participants. The sheer volume of polemical writings and philosophical commentary accompanying the creation of quantum mechanics can be compared, perhaps, with only a case from another culture—the controversy provoked by the first publication of Darwin's *Origins*.[17] Though wider than usual, the circle of those who participated in the interpretational polemics around quantum mechanics was still restricted. It included several recognized leaders, as well as a few unavoidable marginal authors and outsiders to the field, but characteristically not the mainstream contributors to its technical development, the almost 100 postdocs, assistants, and PhD students who authored the majority of publications during the first two years of quantum mechanics. More than twenty years ago I had an opportunity to meet in Göttingen one of the last living members of that cohort, Friedrich Hund, and in the midst of conversation inquired in passing about Hund's own position in the interpretational controversy. He surprised me at the time by replying straightforwardly that it was not his business, but then added, somewhat more expectedly, "but, of course, Bohr was right."[18] As a young assistant in Göttingen and subsequently a postdoctoral fellow in Copenhagen in 1926–27, Hund occupied himself with calculations of molecular spectra using quantum mechanics, but was not entitled to contribute to the public debate about its interpretation.

Besides its strong interest in philosophizing, the academic culture that produced quantum mechanics was also very sensitive to questions of hierarchy, with both these concerns closely linked. After all, contributing to the generally important philosophical outlook was considered the attribute of a truly great scholar, not necessarily of an aspiring or rank and file researcher. In this respect it is somewhat unusual to find among the entitled participants not only ordinary professors and *Geheimräte*, but also Pauli, Heisenberg and Jordan—all extremely important, but still junior contributors to quantum mechanics. Taking a closer look, however, one can see the precariousness of their participation. Pauli was involved mainly in the informal, then unpublished exchange of philosophical ideas, via private correspondence. Jordan essentially entered the public debate on behalf of his professor, Born. And even the recognized pioneer, Heisenberg, before he became a professor himself, had a hard time insisting on his right to publish an interpretation that would become known as the uncertainty principle, violating the existing strict, if unwritten, subordination rules governing publication procedures.

"physicist as philosopher" phenomenon, see (Carson 2010). Hermann von Helmholtz often served as the role model and a typical example of such a combination (Warburg 1922; Krüger 1994).

[17] I am thankful to Simon Schaffer for this observation.

[18] I am grateful to Klaus Hentschel for the invitation to take part in his interview with Hund. See also (Hund 1996).

Participation in the philosophical discourse was thus a mark of prestige, privilege, and status—recognition of not merely the social but also the intellectual hierarchy and a person's crucial contribution to the field. "Perhaps it was also a battle over who did the whole thing first," admitted Heisenberg many years later.[19] In my view, the genre of philosophizing did indeed allow physicists to make claims over the entire theory, but the claims were about property rather than priority. Competing philosophical interpretations did not reorder the chronology of individual contributions to the emerging field, but they reassigned the relative importance of those contributions for "the whole thing." Nobody questioned Heisenberg's credit as the author of quantum mechanics' first proposal, but he was deeply concerned about the decrease in its perceived value during the months when Schrödinger's interpretation rose in popularity. Similarly, nobody tried to or could deprive Schrödinger of his authorship of the theory's central equation, but, depending on the interpretation, his contribution could be labeled primarily as "mathematical" (= technical) rather than "philosophical" (= fundamental). And Bohr, by offering the last, if not final, word on the developing interpretation secured his public reputation as the leader of the new theory, despite the fact that he did not publish on it during its development in 1925–27. In contrast, Schrödinger's failure to establish the prevailing philosophical interpretation signified his loss of control over the field.

The emphasis on what each participant considered his personal major contribution to quantum ideas may explain many of the consistencies and inconsistencies in their philosophical pronouncements. After having invented wave mechanics, Schrödinger abandoned his earlier flirtation with acausality in favor of the (causal) philosophy of continuity and *Anschaulichkeit*. Having reinterpreted the wave function probabilistically, Born and Pauli reversed their pronouncements about acausality and statistics to the affirmative position. Bohr persistently emphasized the fundamentality of discontinuity in quantum phenomena, obviously linked to the postulate of discrete states in his original 1913 model of the atom. Einstein had expressed skepticism about quantum mechanics early on, even before it turned acausal, largely because it did not offer an answer to the crucial question—for him, in view of his earlier contributions to quantum physics—on the wave- or particle-like structure of light (Kojevnikov 2002).

One can imagine a different situation: a major scientific accomplishment belonging, more or less unquestionably, to one distinguished scientist. The ritual of philosophizing would be performed in this case, too, as the privilege and duty of a great scholar, but the leader's right to furnish his theory with a general interpretation would also likely have remained unchallenged. The creation of quantum mechanics, in contrast, was a real group effort, although not a team effort. No other great scientific innovation of the period, including relativity theory, had so many crucial and chronologically close contributions from different authors, each with his own agenda and aspirations, and thus so many potential leaders at once. The existing genre of philosophizing required quantum physicists to translate the meaning of their scientific accomplishment into the language of cultural and ideological values of the

---

[19]Heisenberg, interview by Kuhn (AHQP), quoted in Beller (1996, 556).

time. At the same time, it also offered a culturally approved and respectable form of public discourse within which they could implicitly, and therefore without losing face, debate their rival claims for the entire theory, which inspired them to develop several competing and incompatible translations. The intensity of scientists' philosophical disagreements corresponded to the unusually high level of intra-disciplinary competition; the latter started long before the theory was in any sense completed, as did the former. New and crucial contributions continued coming; even the most basic assumptions of quantum mechanics were still in flux, as well as the relationships between individual authors. Thus also rhetorical strategies kept changing, resulting in opportunistic shifts in announced philosophical principles between 1925 and 1927.

Chapter 7
Conclusions: The Precarious Copenhagen Network and the Forman Thesis

In a 1919 letter to Bohr, his friend and colleague Ehrenfest, professor at the University of Leiden, commented insightfully on postwar changes in the European intellectual climate:

> [I]t is remarkable that precisely here, in the circles of men having much to do with technology, production, industry, patents etc., opinions develop so uniformly about perspectives of culture. Overall there is building up an uncannily intensive reaction *against rationalism*... If I am not entirely mistaken, in the next 5–10 years we will see the following happening at the institutes of higher learning (including technical!). Professors raised as relatively *rational* and disciplined individuals will despairingly and uncomprehendingly face the complaints and demands of a relatively "*mystical*" student body. At the same time, scientifically less clear but personally warmer teachers will gain the main influence over students.[1]

Ehrenfest's testimony remarkably and almost literally supports the core claim of the Forman thesis that, in the immediate wake of World War I, a strong anti-rationalist wave swept through the intellectual public in general, and that even engineers and scientists, professionals who could be expected to strongly resist such trends, started entertaining more mystical lines of thought (Forman 1971). Notably, the letter came from a physicist in the Netherlands to his colleague in Denmark, signifying that the mood did not remain confined to countries that had lost the war, but also affected at least the neighboring neutral lands. Although not prepared to abandon his personal

---

[1]"Und es ist merkwürdig, dass gerade auch hier in den Kreisen der Männer die so viel mit Technik, Industrie, Oekonomie, Teylerstelsel, Patenten etc. zu thun haben, merkwürdig übereinstimmend über die Perspektieven der Kultur urtheilen. Überall bereitet sich eine unheimlich intensive Reaction *gegen den Rationalismus* vor... Wenn ich mich nicht ganz irre wird man in den kommenden 5–10 Jahren an den Hochschulen (auch technischen!) folgendes zu sehen bekommen. Die Professoren werden als relativ *rationalistisch* erzogene und disciplinierte Individuen händeringend und verständnislos den Klagen und Forderungen einer relativ "*mystischen*" Studentenschaft gegenüberstehen. Und wissenschaftlich minder klare aber menschlich wärmere Docenten werden den Haupteinfluss auf die Studenten gewinnen... während ich das niederschreibe ist mir plötzlich viel deutlicher geworden *warum* ich... so sehr viel stärker durch das Uhrtheil der Jüngeren als das der älteren angepackt werde." Ehrenfest to Bohr, 4 June 1919 (emphasis in the original).

A. Kojevnikov, *The Copenhagen Network*, SpringerBriefs in History of Science and Technology, https://doi.org/10.1007/978-3-030-59188-5_7

rationalist convictions, Ehrenfest appeared to defer to the opinions of the younger generation, as academics often do with the newest intellectual fashions.

His letter contained a strikingly self-conscious recognition and expectation that professors would adapt knowingly, rather than unreflectively, to the direction of the prevailing intellectual wind. Ehrenfest's other correspondence reveals quite clearly that, in his own field of theoretical physics, he admired and regarded Bohr as precisely the kind of professor whose charisma would resonate with and inspire students. The quoted description, indeed, sits well with hagiographic recollections that characterize Bohr as a philosophical guru, whose thoughts were too profound to be understood or even expressed clearly, which only helped them to be tremendously inspiring. Whether or not Ehrenfest's letter thus contained an implicit advice, and whether or not Bohr accepted it or arrived at similar ideas on his own, around the same time he was already inclined "to take the most radical *or rather mystical* views imaginable" in thinking about quantum problems.[2]

The preceding chapters analyzed the discourse and practices of physicists during the creation of quantum mechanics in the context of the political and cultural crises of the post-WWI era. Social anxieties and the consequent talk of a "crisis in science" implied not merely the economic difficulties of the profession, but also serious doubts regarding the conceptual basis of existing knowledge. Scientists became much more willing, in comparison with more stable times, to revise or entirely abandon the fundamental principles and commitments of their respective disciplines. In the case of quantum physics, such culturally amplified criticisms were directed not only at the foundational concepts of classical physics, but also at basic ideas of the quantum theory of the atom, which had only been around for a decade but was about to become labeled, unfairly but characteristically, the "old quantum theory."

Value-laden cultural concepts framed the direction of scientists' criticisms, their conceptual vocabulary, and the quest to define new principles. Had a larger share of the debate about quantum phenomena taken place in Great Britain, for example, the question of whether electrons have free will could have acquired more prominence in the new theory. In Central Europe, the philosophical controversy in the emerging quantum mechanics centered around four main issues: *Anschaulichkeit*, quantum discontinuity, the wave–particle dilemma, and causality. Two of them—the first and the last—have been identified by Forman as carrying important and sensitive meanings within the culture of Weimar Germany (Forman 1984). The other two belonged to the general tradition of philosophizing about physics. *Individualität* figured less prominently, but did make an appearance in Bohr's complementarity interpretation, essentially standing in for the indivisibility of quanta. The richness and controversial character of the Weimar cultural field allowed multiple—and not necessarily straightforward ways—for scientists to adapt to it.[3] Philosophical pronouncements that they presented as strongly held principles were in fact often flexible and sometimes mutable to almost their opposites. This complex dynamics of rhetorical shifts

---

[2]Bohr to Darwin, July 1919; see also Bohr to Ehrenfest, 22 October 1919.

[3]For a different proposal of a non-direct and less deterministic adaptation to cultural trends, see Wise (2011).

and changing strategies could be interpreted as justifying competing property claims over the emerging new revolutionary theory at a time when its major concepts were still in flux. The philosophical discourse of quantum physicists utilized various, often rival, and incompatible ways to translate general cultural concerns into the language and problems of their specific field.

Contemporaries overwhelmingly perceived Bohr as the ultimate winner in the debate with Einstein and Schrödinger. Philosophers who analyze the dispute today, however, often find it hard to explain from a logical point of view what made the Copenhagen philosophy preferable to the arguments of its critics. From the criterion of better adaptation to the cultural values of the time, Schrödinger's *Anschaulichkeit* argument also does not seem much weaker than his opponents' acausality claim. How then does one account for the apparent victory of the Copenhagen interpretation in the 1920s? By understanding that the interpretational debate constituted only the most visible tip of the iceberg in the ongoing intradisciplinary rivalry over the new theory. Printed philosophical words by themselves could provide public justifications and rationalizations for the outcome, but did not necessarily decide it. The latter depended more on mainstream contributors to quantum mechanics: fellows, such as Hund, who as a rule did not participate directly in the philosophical polemics, but published the majority of papers and calculations, cited others' works, and together constituted the decisive reference group.

Ehrenfest's 1919 formulation of the Forman thesis *avant la lettre* makes an important addition to it by pointing out the immediate and effective milieu whose demands made professors respond and adapt: the younger generation of students. For Bohr, the main target audience in this regard was not the Danish undergraduates, whom he as a rule did not teach, but more advanced and international doctoral and postdoctoral students. Almost all the junior physicists who worked with him received their degrees elsewhere and usually came to his institute as assistants, temporary visitors, or postdoctoral fellows. Philosophical arguments, mystical overtones, and personal charisma mattered, at least to some of them, but they were also influenced by professional opportunities, available problems to solve, financial considerations, and the institutional authority of their professors. Their movements to and from different centers and the collective body of work submitted for publication from the Copenhagen and Göttingen institutes created the perception of where the leaders of the quantum mechanical revolution were. The ownership of the field in this sense was defined by the international network of precarious and peripatetic contributors.

The construction of this network, with its symbolic center in Copenhagen, was mostly accomplished by 1925. Bohr's astonishing, if counterintuitive, success during the interwar period relied upon his active mobilization of a heterogeneous set of resources—scientific, financial, political, diplomatic, institutional, and rhetorical. The most critical ones included, but were not limited to: his initial pathbreaking contributions to the atomic theory; World War I and the aftermath that divided international science into "hostile political camps"; Denmark's clever use of its political neutrality and offshore status; the international boycott of German and Austrian science that disrupted many existing scholarly networks; postwar European inflation and the hardships it imposed upon researchers, especially junior ones; American financial resources and their culturally specific philanthropy; and last but not least, the ascendance of the international postdoctoral fellow as an established career stage for up-and-coming scientists. These various factors and unique historical circumstances combined to allow Bohr an opportunity to gradually build up, starting from an extremely modest base, the Copenhagen Institute for Theoretical Physics and its global network of quantum physicists.

The story thus construed presents Bohr as a great politician, diplomat, fundraiser, and manager of research. These full-time commitments and crucial functions have often been underappreciated or obscured in the traditional disciplinary myth, which depicted Bohr's primary role and influence as that of a philosophical sage who guided research questions and mathematical efforts of younger fellows. The unique scholarly and social network that he developed produced, by the end of 1927, a great intellectual revolution, quantum mechanics. This landmark scientific breakthrough became possible, in no small measure, because the institutional success in extending the network at some point outgrew senior professors' capacity to effectively control it, thus allowing at least some of its junior participants additional degrees of intellectual freedom to pursue radical ideas. Transitory students who constituted the majority of contributors acquired a sufficient critical mass to develop their own, postdoctoral culture with its specific *modus operandi* and collective research momentum. Quantum mechanics can thus be properly characterized as resulting from the international "postdoctoral revolution" in science, or as *Knabenphysik* without irony.

The unique constellation of circumstances that enabled the quantum mechanics story ultimately could not withstand the more powerful and dangerous historical whirlwinds of the subsequent decades. The photo of the 1936 conference in Copenhagen looks seemingly normal, just like a dozen other annual conferences in the same room with many of the same participants sitting together and projecting the impression of a scholarly community still functioning in the same collegial mode as before. Yet the fragile and localized internationalism of the 1920s had already been destroyed by that time. Without additional knowledge, you would not be able to tell from this photo who among the participants had already become a refugee, driven away by anti-Semitic laws; who was still trying to hang on to a job but would soon lose it; who fled to Copenhagen as a temporary shelter on the way toward permanent emigration; who had joined the Nazi party or stayed in Germany to face inevitable and difficult moral and political compromises; who from the earlier participants was no longer present, having committed a desperate suicide, or now forbidden from traveling abroad; who was relatively safe on neutral soil; and who would soon live under the dangers of military occupation.[4] But one way or another, the former quantum mechanical community was about to be split, politically and personally. Many of its members would become involved in nuclear research for different governments and opposing sides, and would later feel responsible, if indirectly, for the eventual development of horrible weapons (Frayn 1998, Lemmerich 2011, Carson 2010). But for

---

[4]See Hoffmann (1988) for the political analysis of controversies and compromises related to the 1936 conference.

Fig. 7.1 Annual conference at Bohr's institute, June 1936

the rest of their lives, they continued cherishing, as the brightest and most nostalgic memory, their shared, unrepeatable experience of partaking in something as exciting and profound as the birth of quantum mechanics (Fig. 7.1).

Appendix A
Visitors at Bohr's Institute in Copenhagen (Chronological Order, up to 1927)

Start	Name	Country	End	Funding
1916, September	H.A. Kramers	Netherlands	1926, May	CF
1918, May	O. Klein	Sweden	1922, June	RØF, CF
1919, September	A. Sommerfeld	Germany	1919, September	DK
1920, March	A. Rubinowicz	Poland	1920, August	RØF
1920, March	G. de Hevesy	Hungary	1926, September	RØF
1920, October	A. Landé	Germany	1920, October	RØF
1920, September	E. Rutherford	UK	1920, September	DK
1920, September	S. Rosseland	Norway	1924, August	RØF, CF
1921, January	J. Franck	Germany	1921, March	DK
1921, March	T. Takamine	Japan	1921, July	
1921, August	A. Udden	USA	1922, August	SAF
1921, December	P. Ehrenfest	Netherlands	1921, December	
1922, May	A. Rubinowicz	Poland	1922, May	RØF
1922, June	B. Lindsay	USA	1923, June	SAF
1922, September	D. Coster	Netherlands	1923, September	RØF
1922, October	W. Pauli	Austria	1923, October	RØF
1922, October	F. C. Hoyt	USA	1924, September	NRF, RØF
1923, April	Y. Nishina	Japan	1928, October	UTS, RØF
1923, September	H. C. Urey	USA	1924, June	SAF
1923, December	J. C. Slater	USA	1924, April	STF
1923, December	E. R. Jette	USA	1924, June	SAF
1924, March	F. Paschen	Germany	1924, March	DK
1924, March	W. Kuhn	Switzerland	1926, February	IEB
1924, August	V. M. Goldschmidt	Norway	1924, August	
1924, September	W. Heisenberg	Germany	1925, April	IEB

(continued)

© The Author(s), under exclusive license to Springer Nature Switzerland AG 2020
A. Kojevnikov, *The Copenhagen Network*, SpringerBriefs in History of Science
and Technology, https://doi.org/10.1007/978-3-030-59188-5

(continued)

Start	Name	Country	End	Funding
1924, September	L. Ebert	Germany	1925, August	IEB
1924, October	A. Obrutscheva	USSR	1925, April	
1924, October	B. B. Ray	India	1925, September	
1924, October	D. M. Dennison	USA	1926, June	IEB
1925, January	R. de L. Kronig	USA	1925, November	BCF
1925, February	R. H. Fowler	UK	1925, April	IEB
1925, February	M. Born	Germany	1925, February	DK
1925, April	W. Pauli	Germany	1925, April	
1925, April	T. Takamine	Japan	1925, November	
1925, May	K. Kimura	Japan	1927, April	UTS
1925, July	I. Waller	Sweden	1926, June	
1925, September	W. Heisenberg	Germany	1925, October	IEB
1925, September	J. H. Dewey-Clark	USA	1927, February	BF, RØF
1925, October	L. H. Thomas	UK	1926, June	INS
1925, November	M. Y. Sugiura	Japan	1927, April	UTS
1926, February	S. A. Goudsmit	Netherlands	1926, February	
1926, March	O. Klein	Sweden	1931, January	RØF, CF
1926, April	C. G. Bedreag	Rumania	1926, June	
1926, May	W. Heisenberg	Germany	1927, June	CF
1926, September	J. S. Foster	Canada	1927, January	IEB
1926, September	S. Aoyama	Japan	1927, March	UTS
1926, September	P. A. M. Dirac	UK	1927, February	ES, DSIR
1926, September	F. Hund	Germany	1927, April	IEB
1926, September	T. Hori	Japan	1927, April	UTS
1926, September	S. Rosseland	Norway	1927, April	RØF, CF
1926, October	E. Schrödinger	Switzerland	1926, October	DK
1927, January	E. Fues	Germany	1927, April	IEB
1927, March	D. M. Dennison	USA	1927, April	
1927, March	C. G. Darwin	UK	1927, June	
1927, March	E. H. Kennard	USA	1927, June	
1927, April	S. A. Goudsmit	Netherlands	1927, June	IEB
1927, April	J. P. Holtsmark	Norway	1927, July	
1927, April	E. Hulthén	Sweden	1929, February	RØF
1927, April	G. E. Uhlenbeck	Netherlands	1927, June	LF
1927, April	I. Waller	Sweden	1927, June	
1927, April	L. Pauling	USA	1927, June	GF
1927, May	O. Richardson	UK	1927, May	DK

(continued)

(continued)

Start	Name	Country	End	Funding
1927, May	H. Faxén	Sweden	1927, July	
1927, May	P. Jordan	Germany	1927, November	IEB
1927, June	R. de L. Kronig	USA	1927, December	IEB
1927, June	G. Wentzel	Germany	1927, June	
1927, September	I. I. Rabi	USA	1927, October	BF

Sources: Robertson (1979, 156–159), Marner (1997, 95–100), NBA; RAC. Some dates are approximate
Funding (when known): DK (Danish), including CF (Carlsberg Fond), RØF (Rask-Ørsted Fond)
Home country: including SAF (Scandinavian-American Foundation, USA); UTS (University of Tokyo Scholarship, Japan); BF (Barnard Fellowship, USA); DSIR (Department of Science and Industrial Research, UK); BCF (Bayard-Cutting Travelling Fellowship, USA); GF (Guggenheim Fellowship, USA); INS (Isaac Newton Student, UK); NRF (National Research Fellowship, USA); STF (Sheldon Travelling Fellowship, USA); ES (1851 Exhibition Scholarship, UK); LF (Lorentz Fellowship, Netherlands)
IEB (International Education Board)

Appendix B

Information Concerning Fellowships in Science Awarded by the International Education Board (1925)

1—A limited number of fellowships will be granted by the International Education Board to assist young scientific men who are working under the direction of scientists. Grants will be made to men of unusual promise in their fields, so that they may pursue abroad, under guidance, studies which they cannot pursue at home. In making grants, the board will give preference, for the present, to men under thirty-five years of age who are working in the fields of chemistry, biology, physics, and mathematics. An adequate knowledge of the language of the country which they intend to visit is required. The grant will cover the period needed for the contemplated course of study abroad, provided it is not less than six months. In case more than a year is required, the fellowship will be granted for the year, and a request for an extension will be considered by the board. It is the purpose of these fellowships to promote an exchange of professional experience on an international scale...

3—Application for Fellowships. Applications are submitted to the board, not by the beneficiary of the fellowship, but by scientists who are personally acquainted with the character and quality of his work, and who are prepared to act as his sponsors. They must take the initiative, in substance as well as in form: and the board assumes that their nominations will be made with a full sense of professional responsibility.

To assist in furnishing information desired by the board, a personal history blank is provided. These blanks may be obtained from the board, on behalf of a qualified young scientist, by the man under whose direction he is pursuing his scientific work. These record and history, supplied by the candidate, are returned to his superior, who if he approves, signs under the title "proposed," and forwards it to the person abroad under whom the work is contemplated. If the second sponsor approves, he signs under the title "seconded" and returns the blank to the proposer. The proposer then forwards the document to the International Education Board, 61 Broadway, New York City, accompanied by a formal application for a fellowship grant.

The board desires to receive from each fellow, at the close of his work, a statement covering his activities during the fellowship period.

[RAC. IEB. 1.3.42.599].

A. Kojevnikov, *The Copenhagen Network*, SpringerBriefs in History of Science and Technology, https://doi.org/10.1007/978-3-030-59188-5

Commentary: Over several years, the IEB and the Rockefeller Foundation issued somewhat changing formulations of the above fellowship rules. The application procedures were not always followed literally but allowed some flexibility. In particular, despite the strictly sexist language of the document, a small number of fellowships were also awarded to female scientists.

Appendix C
IEB: Excerpts from Applications and Correspondence Regarding Individual Fellows

Bohr to August Trowbridge, IEB, 2 March 1925 [RAC. RF. 6.1. 1.1. 9. 87].

"Dr. Heisenberg, who has now been here for half a year is, as I need hardly say, a most promising and ingenious young physicist. During his stay here he has made several important contributions to the problems of radiation which in these years is a main topic for the work in this institute. Thus he has in this autumn written a very illuminating note on the problem of intensities of spectral lines and polarisation of resonance radiation which will appear soon in Zeitschrift für Physik. At the same place there will also soon appear a larger communication of Kramers and Heisenberg in which some new and interesting features of the effect of atoms in scattering of radiation are discussed. At present he is occupied with the problems of the anomalous Zeeman effect and complex structure of spectral lines, on which problems it seems that the new light might be thrown by means of the notions of the symbolic nature of energy and momentum, brought out by the discussion of the interaction between matter and radiation, and to which quite recently Kramers has made an, I believe, very important contribution. For some of the early summer months Heisenberg will have to go back to Göttingen to perform some lecturing duties, but in order to work here for a full year, corresponding with his stipend, he intends to come back for the autumn months of September and October when there is vacation in Göttingen. He wishes himself to stay on for the next year and we would like very much indeed to keep him, but on account of his future settling down in Germany he is advised not to stay away for more than a year at the time, especially since he has been created Privatdozent in Göttingen just before he came here. I hope, however, that he shall be able to come back another time and I should be glad by occasion to hear your opinion whether he might obtain a fellowship for another year even if it is not in direct prolongation of his present fellowship."

Trowbridge to William W. Bierley, IEB, 10 March 1926 [RAC. IEB. 13. 52. 808].

"I beg to inform you that in the case of the fellowship granted to Doctor FRIEDRICH HUND, which was to begin on or about March 1st 1926 for a period of study of about 6 months to permit him to go to Denmark, I have been requested by Professor Max Born, whose Assistant at Göttingen Dr. Hund is, to authorize a postponement until the 1st of September 1926 for the following reasons. Professor Born's former Assistant, Dr. Heisenberg, who was also a fellow of the Board, has been called to an assistantship at Copenhagen with Professor Niels Bohr and has accepted. When the application was made for Dr. Hund, who is filling the vacant assistantship of Dr. Heisenberg, it was thought that Heisenberg would return about the time that Hund left, if awarded a fellowship. Professor Born asks for a delay, in order that he may not be left without any assistant until he can make new arrangements, which he expects he will be able to do by the end of the summer. This change of plan seems to me wholly in the interests of science and I have, therefore, in the name of the Board, authorized the change."

Schrödinger to Trowbridge, IEB, 29 April 1926 [RAC. IEB. 1.3 49. 739].
"I am greatly enjoying the presence of Mr. E. Fues here in Zurich, which I owe to the generosity of the International Education Board. It was for the purposes of the Board a very favourable circumstance that I was lucky enough to detect a new quantum-theory in the beginning of this year, which may perhaps be thought of as the beginning of **the** solution of quantum-difficulties. My hopes are much encouraged by Mr. Sommerfeld, Mr. Planck and Mr. Einstein, the latter wrote me a few days ago, that he thought my solution better than the one proposed at Göttingen (Heisenberg, Born, Jordan).

Mr. Fues has helped me a great deal in this work, he has just now sent to Annalen der Physik a paper on the theory of bandspectra according to the new conceptions. He is continuing this work and will, I hope, be able to publish another paper in the course of this year on the most interesting subject of **intensities** in bandspectra. As I unfortunately have nearly **no** real pupils here in Zurich, I come to think with some despair of the time—happily still somewhat remote—when Mr. Fues will return to Stuttgart. I should like to ask you, Mr. Trowbridge, whether you think it feasible, that another fellowship would be granted to some young men to come to Zurich to study and work on the new theory."

David M. Dennison to W.E. Tisdale, IEB, 26 October 1926 [RAC. IEB. 1. 47. 695].
"I arrived in Copenhagen at the end of September 1924. During the first months I studied the periodicals under Professor Bohr's direction. After this I made a theoretical analysis of certain molecular spectra and was able to determine the forms and force functions for a number of molecules. (HCl. HBr. CO. CO_2 and NH_3) This work extended over most of the winter and was completed in June 1925. It has been published in the Philosophical Magazine…

My second year opened with a study of the infra red spectrum of water vapor. Very suggestive results were obtained which I intend to continue later but the work was broken off in order to study the remarkably important work of Heisenberg on Quantum Mechanics. In November 1925 I commenced to work on an application of his ideas to certain molecular rotators. Equations were found which permitted systems such as these which involve restraints to be treated. I obtained the theoretical energies, frequencies and intensity amplitudes of (1) the simple rotator in a plane, (2) the simple rotator in space and (3) molecules having an axis of symmetry (A = B, C). Experimental data are not yet in existence to verify the theoretical results but there will be no principal difficulty in obtaining them. The paper appeared in the Physical Review for August 1926..."

Personal History Record Submitted in Connection with Application for a Fellowship. 15 January 1927 [RAC. IEB. 1. 53. 839. Kronig].

Name in full: Ralph de Laer Kronig. Present position: Lecturer, Columbia University.

Place of birth: Dresden. Date of birth: March 10, 1904. Citizenship: USA.

Describe in detail the study or investigation you wish to carry on: The new quantum mechanics of Born, Heisenberg, and Schrödinger does not as yet take into account what corresponds to the radiation reaction in the classical theory. An extension of the theory so as to include a radiation reaction is at present a most urgent problem. It must account especially for the coupling pairs of the emission and absorption processes.

Why do you wish to carry on the study or investigation above named at the place you have chosen? Primarily because the great concentration of theoretical physicists working there in the field of atomic structure gives an opportunity of discussing all questions that may arise and provides the "theoretical atmosphere" so essential for good work. Secondly one learns of any progress made in the theory months ahead of the appearance of the publication.

What are your plans following completion of fellowship: to resume an academic position (probably at Columbia University)."

Bohr to Tisdale, IEB, 26 January 1927 [RAC. IEB. 13. 49. 739].

"Dr. Foster from Montreal, who has been here since last summer and is just on the point of leaving, has been primarily occupied with the theoretical analysis of his very beautiful experimental results concerning the Stark effect, which he brought with him from McGill. On this problem he has obtained most interesting and important results, having succeeded in the new quantum mechanics of Heisenberg to explain almost every detail of his Stark effect patterns. On the whole it may be said that this agreement between experiment and calculation offers a most striking confirmation of the new theory. Although mainly an experimentator by training Dr. Foster has thrown himself with unusual energy into this rather subtle theoretical work and in spite of the shortness of his stay acquired an insight which, I am sure, will prove useful in his continued experimental researches."

Tisdale to Wickliffe Rose, IEB, 1 March 1927 [RAC. IEB. 1. 3. 52. 819].

"Bohr is willing to receive Dr. Jordan and work with him during the summer. The candidate proposes to work on new developments in theoretical researches on quantum mechanics... The application is made for six months for the reason that Prof. Born is engaged in some very serious researches on the new development of quantum mechanics and feels that he cannot spare Jordan after November 1927, but both he and Prof. Bohr think that Dr. Jordan, for his own sake, should spend six months with

Prof. Bohr in Copenhagen. On my recent visit to Germany I talked with the candidate who has a very imperceptible impediment in his speech but is otherwise very keen, attractive and unusually intelligent. I think he is head and shoulders above any of the Americans at Göttingen. Prof. Born and Prof. Franck, upon whose judgment I would rely, inform me that Dr. Jordan is perhaps the most outstanding man of his age in Germany today; they consider him to be of the type of Heisenberg and his accomplishments furnish very good evidence of his promise for the future."

References

Aaserud F (1990) Redirecting science: Niels Bohr, philanthropy, and the rise of nuclear physics. Cambridge University Press, Cambridge

Aaserud F, Heilbron John L (2013) Love, literature, and the quantum atom: Niels Bohr's 1913 trilogy revisited. Oxford University Press, Oxford

Assmus A (1993) The creation of postdoctoral fellowships and the siting of American scientific research. Minerva 31:151–183

Banerjee S (2016) Transnational quantum: quantum physics in India through the lens of Satyendranath Bose. Phys Perspect 18:157–181

Beller M (1983) Matrix theory before Schrödinger: problems, philosophy, consequences. Isis 74:469–491

Beller M (1990) Born's probabilistic interpretation: a case study of 'concepts in flux'. Stud His Philos Sci 21:563–588

Beller M (1992) Schrödinger's dialogue with Göttingen-Copenhagen physicists. In: Bitbol M, Darrigol O (eds) Erwin Schrödinger: philosophy and the birth of quantum mechanics. Editions Frontières, Gif-sur-Yvette, France, pp 277–306

Beller M (1996) The conceptual and anecdotal history of quantum mechanics. Found Phys 26:545–557

Beller M (1999) Quantum dialogue: the making of a revolution. The University of Chicago Press, Chicago

Ben-David J (1971) The scientist's role in society: a comparative study. Prentice-Hall, Englewood Cliffs, NJ

Birks JB (1963) Rutherford at Manchester. Benjamin, New York

Bohr N (1913) On the constitution of atoms and molecules (part I). Philos Mag 26:1–25

Bohr N (1918) On the quantum theory of line-spectra, part I. On the general theory. Kgl. Danske Vid. Selsk. Skrifter IV.1:1–36

Bohr N (1921) Atomic structure. Nature 107:104–107; 108:208–209

Bohr N (1922) Drei Aufsätze über Spektren und Atombau. Vieweg, Braunschweig

Bohr N (1923) Instituttet for teoretisk Fysik. In (Munch-Petersen 1923), 316–329

Bohr N (1924) Zur Polarisation des Fluoreszenzlichtes. NW 12: 1115–1117

Bohr N (1925) Atomic theory and mechanics. Nature 116:845–852

Bohr N (1926) Spinning electrons and the structure of spectra. Nature 117:265

Bohr N (1928) The quantum postulate and the recent development of atomic theory. In: Atti del Congresso Internazionale dei Fisici, vol 2. Nicola Zanichelli, Bologna, pp 565–588

Bohr N (1972–2008) Collected works (13 vols). North-Holland, Amsterdam (cited as BCW)

Bohr N, Kramers HA, Slater JC (1924) The quantum theory of radiation. Philos Mag 47:785–822

Born M (1924) Über Quantenmechanik. ZP 26:379–395
Born M (1926a) Problems of atomic dynamics. MIT Press, Cambridge, MA
Born M (1926b) Zur Quantenmechanik der Stossvorgänge. ZP 37:863–867
Born M (1927) Quantenmechanik und Statistik. NW 5:238–242
Born M (1928) Sommerfeld als Begründer einer Schule. NW 16:1035–1036
Born M (1978) My life: recollections of a Nobel laureate. Scribner, New York
Born M, Heisenberg W (1923) Die Elektronenbahnen im angeregten Helium-atom. ZP 16:229–243
Born M, Jordan P (1925) Zur Quantenmechanik. ZP 34:858–888
Born M, Heisenberg W, Jordan P (1926) Zur Quantenmechanik. II. ZP 35:557–615
Bose SN (1924) Plancks Gesetz und Lichtquantumhypothese. ZP 26:176–181
Bricmont J (2016) Making sense of quantum mechanics. Springer International
de Broglie L (1924) Recherche sur la théorie des quanta. Masson et Cie, Paris
Cahan D (1985) The institutional revolution in German physics, 1865–1914. HSPS 15:1–65
Carlsbergfondets Direktion (ed) (1930) Carlsbergfondet, 1876–1926: et Jubilæumsskrift. 2 vols.
    Lunos Bogtrykkeri, København
Carson C (2010) Heisenberg in the atomic age: science and the public sphere. Cambridge University
    Press, Cambridge
Carson C, Kojevnikov A, Trischler H (eds) (2011) Weimar culture and quantum mechanics: selected
    papers by Paul Forman and contemporary perspectives on the Forman thesis. World Scientific,
    Singapore
Cassidy DC (1979) Heisenberg's first core model of the atom: the formation of a professional style.
    HSPS 10:187–224
Cassidy DC (1992) Uncertainty. The life and science of Werner Heisenberg. Freeman, New York
Cock AG (1983) Chauvinism and internationalism in science: the international research council,
    1919–1926. Notes Rec R Soc 37:249–288
Coster D, Hevesy G (1923) On the missing element of atomic number 72. Nature 111:79
Crawford E (1992) Nationalism and internationalism in science, 1880–1939. Four studies of the
    Nobel population. Cambridge University Press, Cambridge
Cushing JT (1994) Quantum mechanics: historical contingency and the Copenhagen hegemony.
    University of Chicago Press, Chicago
Dahms H-J (2002) Appointment politics and the rise of modern theoretical physics at Göttingen.
    In: Rupke N (ed) Göttingen and the development of the natural sciences. Wallstein, Göttingen,
    pp 143–157
Darrigol O (1992) From c-numbers to q-numbers: the classical analogy in the history of quantum
    theory. University of California Press, Berkeley
Darrigol O (1993) Strangeness and soundness in Louis de Broglie's early works. Physis 30:303–372
Davies SM (1985) American physicists abroad: Copenhagen, 1920–1940. PhD diss, University of
    Texas, Austin
Desser M (1991) Zwischen Skylla und Charybdis: Die "Scientific Community" der Physiker, 1919–
    1939. Böhlau, Vienna
Dirac PAM (1927a) The physical interpretation of the quantum dynamics. Proc Royal Soc Lond A
    113:621–641
Dirac PAM (1927b) The quantum theory of the emission and absorption of radiation. Proc Royal
    Soc Lond A 114:243–265
Dresden M (1987) H. A. Kramers: between tradition and revolution. Springer, New York
Duncan A, Janssen M (2019) Constructing quantum mechanics. Volume 1: the scaffold, 1900–1923.
    Oxford University Press, New York
Eckert M (1993) Die Atomphysiker: Eine Geschichte der theoretischen Physik am Beispiel der
    Sommerfeldschule. Vieweg, Braunschweig
Ehlers J, Hoffmann D, Renn J (eds) (2007) Pascual Jordan (1902–1980): Mainzer Symposium zum
    100. Geburtstag. Preprint MPIWG, No. 329
Einstein A (1916) Strahlungs-Emission und -Absorption nach der Quantentheorie. Verhandlungen
    der Deutschen physikalischen Gesellschaft. 18:318–323

Einstein A (1924–25) Quantentheorie des einatomigen idealen Gases. Sitzungsberichte der Königlich Preussischen Akademie der Wissenschaften zu Berlin (1924) 261–267; 2. (1925) 3–14

Enz CP (2006) No time to be brief: a scientific biography of Wolfgang Pauli. Oxford University Press, Oxford

Epstein PS (1916) Zur Theorie des Starkeffekts. PZ 17:148–150

Eve AS (1939) Rutherford: Being the Life and Letters of the Rt. Hon. Lord Rutherford, O.M. Cambridge University Press, Cambridge

Fick HD, Kant H (2008) Walther Bothe's contributions to the particle-wave dualism of light. Preprint MPIWG, No. 360

Forman P (1967) The environment and practice of atomic physics in Weimar Germany: a study in the history of science. PhD diss. University of California, Berkeley

Forman P (1969) The discovery of the diffraction of X-rays by crystals: a critique of the myth. Arch His Exact Sci 6:38–71

Forman P (1970) Alfred Landé and the anomalous Zeeman effect, 1919–1921. HSPS 2:153–261

Forman P (1971) Weimar culture, causality, and quantum theory, 1918–1927: adaptation by German physicists and mathematicians to a hostile intellectual environment. HSPS 3:1–115

Forman P (1973) Scientific internationalism and the Weimar physicists: the ideology and its manipulation in Germany after World War I. Isis 64:151–180

Forman P (1974) The financial support and political alignment of physicists in Weimar Germany. Minerva 12:39–66

Forman P (1984) Kausalität, Anschaulichkeit, and Individualität, or how cultural values prescribed the character and the lessons ascribed to quantum mechanics. In: Stehr N, Meja V (eds) Society and knowledge: contemporary perspectives in the sociology of knowledge and science. Transaction Books, New Brunswick, NJ, pp 333–347

Franck J, Hertz G (1919) Die Bestätigung der Bohrschen Atomtheorie im optischen Spektrum durch Untersuchung der unelastischen Zusammenstösse langsamer Elektronen mit Gasmolekülen. PZ 20:132–143

Frayn M (1998) Copenhagen. Anchor Books, New York

Friedman RM (1989) Text, context, and quicksand: method and understanding in studying the Nobel science prizes. HSPBS 20:63–77

Friedman RM (1990) The Nobel Prizes and the invigoration of Swedish science: some considerations. In: Tore Frängsmyr (ed) Solomon's House revisited: the organization and institutionalization of science: Nobel symposium 75. Science History Publications, Canton, MA, pp 193–207

Friedman RM (2011) The politics of excellence: behind the Nobel Prize in science. Freeman Book, New York

Gearhart CA (2014) The Franck-Hertz experiments, 1911–1914: experimentalists in search of a theory. Phys Perspect 16:293–343

Glamann K (1976) Carlsbergfondet. Rhodos, København

Glamann K (2002) Jacobsen of Carlsberg: Brewer and Philanthropist. Gyldendal, Copenhagen

Greenspan N (2005) The end of the certain world: the life and science of Max Born. Basic Books, New York

Heilbron JL (1967) The Kossel-Sommerfeld theory and the ring atom. Isis 58:451–485

Heilbron JL (1977) Lectures on the history of atomic physics, 1900–1922. In: Weiner C (ed) History of twentieth century physics. Academic Press, New York, pp 40–108

Heilbron JL (1985) The earliest missionaries of the Copenhagen spirit. Revue d'histoire des sciences 38:195–230

Heilbron JL, Kuhn TS (1969) The genesis of the Bohr atom. HSPS 1:211–290

Heisenberg W (1922) Zur Quantentheorie der Linienstruktur und der anomalen Zeemaneffekte. ZP 8:273–297

Heisenberg W (1925a) Über eine Anwendung des Korrespondenzprinzips auf die Frage nach der Polarisation des Fluoreszenzlichtes. ZP 31:617–628

Heisenberg W (1925b) Zur Quantentheorie der Multiplettstruktur und der anomalen Zeemaneffekte. ZP 32:841–860

Heisenberg W (1925c) Über die quantentheoretische Umdeutung kinematischer und mechanischer Beziehungen. ZP 33:879–893

Heisenberg W (1926) Über quantentheoretische Kinematik und Mechanik. Mathematische Annalen 95:683–705

Heisenberg W (1927) Über den anschaulichen Inhalt der quantentheoretischen Kinematik und Mechanik. ZP 43:172–198

Heisenberg W (1971) Physics and beyond: encounters and conversations. Harper & Row, New York

Heisenberg W (2003) Hirsch-Heisenberg AM (ed) Liebe Eltern! Briefe aus kritischer Zeit 1918 bis August 1945. Langen Müller, Munich

Hendry J (1984) The creation of quantum mechanics and the Bohr-Pauli dialogue. Reidel, Dordrecht

Hentschel K (2018) Photon: the history and mental models of light quanta. Springer International

Hochadel O (2008) The sale of shocks and sparks: Itinerant electricians in the German enlightenment. In: Bensaude-Vincent B, Blondel C (eds) Science and spectacle in the European enlightenment. Ashgate, Burlington, VT, pp 89–101

Hoffmann D (1988) Zur Teilnahme deutscher Physiker an den Kopenhagener Physikerkonferenzen nach 1933 sowie am 2. Kongreß für Einheit der Wissenschaft, Kopenhagen 1936. NTM 25:49–55

Hon G (1989) Franck and Hertz versus Townsend: a study of two types of experimental error. HSPBS 20:79–106

Howard D (1994) What makes a classical concept classical? Toward a reconstruction of Niels Bohr's philosophy of physics. In: Faye J, Folse HJ (eds) Niels Bohr and contemporary philosophy. Kluwer, Dordrecht, pp 201–229

Hund F (1974) The history of quantum theory. Barnes and Noble, New York

Hund F (1985) Bohr, Göttingen, and quantum mechanics. In: French AP, Kennedy PJ (eds) Niels Bohr. A centenary volume. Harvard University Press, Cambridge MA, pp 71–75

Hund F (1987) Geschichte der Physik in Göttingen. Vandenhoeck & Ruprecht, Göttingen

Hund F (1996) Friedrich Hund zum 100. Geburtstag, befragt von Klaus Hentschel und Renate Tobies. NTM 4:1–18

Jammer M (1966) The conceptual development of quantum mechanics. McGraw-Hill, New York

Johnson JA (2017) Between nationalism and internationalism. The German chemical society in comparative perspective, 1867–1945. Angewandte Chemie 56:11044–11058

Jordan P (1927a) Über eine neue Begründung der Quantenmechanik. ZP 40:809–838

Jordan P (1927b) Kausalität und Statistik in der modernen Physik. NW 15:105–110

Jungnickel C, McCormmach R (1986) Intellectual mastery of nature: theoretical physics from Ohm to Einstein (2 vols). University of Chicago Press, Chicago (cited as IMN).

Kaiserfeld T (1993) In: Lindqvist (ed) When theory addresses experiment: the Siegbahn-Sommerfeld correspondence, 1917–1940, pp 306–324

Kevles DJ (1971a) "Into hostile political camps": The reorganization of international science in World War I. Isis 62:47–60

Kevles DJ (1971b) The physicists: the history of a scientific community in modern America. Harvard University Press, Cambridge, MA

Klein O, Rosseland S (1921) Über Zusammenstöße zwischen Atomen und freien Elektronen. ZP 4:46–51

Knudsen H, Nielsen H (2012) In: Letteval et al (eds) Pursuing common cultural ideals: Niels Bohr, neutrality, and international scientific collaboration during the interwar period, pp 115–139

Kojevnikov A (2002) Einstein's fluctuation formula and the wave-particle duality. Einstein Studies 10:181–228

Kojevnikov A (2011) In: Carson et al (eds) Philosophical rhetoric in early quantum mechanics 1925–27: high principles, cultural values and professional anxieties, pp 319–348

Kormos Barkan D (1992) A usable past: creating disciplinary space for physical chemistry. In: Nye MJ, Richards JL, Stuewer RH (eds) The invention of physical science. Springer, Dordrecht, pp 175–202

Kozhevnikov A, Novik O (1989) Analysis of informational ties in early quantum mechanics (1925–1927). Acta Historiae Rerum Naturalium necnon Technicarum 20:115–159

Kragh H (1979) Niels Bohr's second atomic theory. HSPS 10:123–186

Kragh H (1980) Anatomy of a priority conflict: the case of element 72. Centaurus 23:275–301

Kragh H (1985) The fine structure of hydrogen and the gross structure of the physics community, 1916–1926. HSPS 15:67–125

Kragh H (2011) Resisting the Bohr atom: the early British opposition. Phys Perspect 13:4–35

Kragh H (2012) Niels Bohr and the quantum atom: the Bohr model of atomic structure, 1913–1925. Oxford University Press, Oxford

Kramers HA (1919) Intensities of spectral lines: on the application of the quantum theory to the problem of the relative intensities of the components of the fine structure and of the Stark effect of these lines of the hydrogen spectrum. Kgl. Danske Vidensk. Selsk. Skrifter III 3:284–386

Kramers HA (1923) Über das Modell des Heliumatoms. ZP 13:312–341

Kramers HA, Heisenberg W (1925) Über die Streuung von Strahlung durch Atome. ZP 31:681–708

Kramers HA, Pauli W (1923) Zur Theorie der Bandenspektren. ZP 13:351–367

Krüger L (ed) (1994) Universalgenie Helmholtz. Rückblick nach 100 Jahren. Akademie-Verlag, Berlin

Latour B (2005) Reassembling the social: an introduction to actor-network-theory. Oxford University Press, Oxford

Lemmerich J (2011) Science and conscience: the life of James Franck. Stanford University Press, Stanford

Lettevall R, Somsen G, Widmalm S (eds) (2012) Neutrality in twentieth-century Europe: intersections of science, culture, and politics after the First World War. Routledge, New York

Levi H (1985) George de Hevesy. Life and work. Rhodos, Copenhagen

Lindqvist S (ed) (1993) Center on the periphery: historical aspects of 20th-century Swedish physics. Science History Publications, Canton, MA

Mackinnon E (1977) Heisenberg, models, and the rise of matrix mechanics. HSPS 8:137–188

Marner J (1997) Bohr's Institute in Copenhagen and the interpretation of quantum mechanics in the Weimar period. Cand. Sci. thesis. The Niels Bohr Institute, Copenhagen

Mehra J, Rechenberg H (1982–2001) The historical development of quantum theory (6 vols). Springer, New York (cited as HDQT)

von Meyenn K, Stolzenburg K, Sexl RU (eds) (1985) Der Kopenhagener Geist in der Physik. Vieweg, Braunschweig

Moore W (1989) Schrödinger: life and thought. Cambridge University Press, Cambridge

Munch-Petersen H (ed) (1923–1925) Aarbog for Københavns Universitet, Kommunitetet og Polytekniske Læreanstalt indeholdende Meddelelser for de akademiske aar 1915–1920. Parts III–IV. Københavns Universitet, København

Murdoch D (1987) Niels Bohr's philosophy of physics. Cambridge University Press, Cambridge

Nielsen JR (1963) Memories of Niels Bohr. Phys Today 16(10):22–30

Ostwald W (1926–1927) Lebenslinien. Eine Selbstbiographie (3 vols). Klasing, Berlin

Pais A (1991) Niels Bohr's times in physics, philosophy, and polity. Clarendon, Oxford

Pauli W (1923) Über das thermische Gleichgewicht zwischen Strahlung und freien Elektronen. ZP 18:272–286

Pauli W (1925) Über den Zusammenhang des Abschlusses der Elektronengruppen im Atom mit der Komplexstruktur der Spektren. ZP 31:765–783

Pauli W (1979) Hermann A, Meyenn KV, Weisskopf VF (eds) Wissenschaftlicher Briefwechsel mit Bohr, Einstein, Heisenberg u. a. (vol 1: 1919-1929). Springer, New York (cited as PWB)

Pihl M (1983) Fysik. In: Københavns Universitetet, 1479–1979 (vol 12). G.E.C. Gads Forlag, København, pp 365–426

Planck M (1975) Laudatio für A. Einstein. In: Kirsten Ch, Körber H-G (eds) Physiker über Physiker (vol 1). AkademieVerlag, Berlin, p 202

Radder H (1982) Between Bohr's atomic theory and Heisenberg's matrix mechanics. A study of the Dutch physicist H. A. Kramers. Janus 69:223–252

Robertson P (1979) The early years: the Niels Bohr Institute, 1921–1930. Akademisk Forlag, Copenhagen

Rubinowicz A (1918) Bohrsche Frequenzbedingung und Erhaltung des Impulsmomentes. PZ 19:441–445

Scerri ER (1994) Prediction of the nature of hafnium from chemistry, Bohr's theory, and quantum theory. Ann Sci 51:137–150

Schirrmacher A (2019) Establishing quantum physics in Göttingen: David Hilbert, Max Born, and Peter Debye in context, 1900–1926. Springer, Berlin

Schröder-Gudehus B (1978) Les scientifiques et la paix: La communauté scientifique internationale au cours des années 20. Presses de l'Université de Montréal, Montreal

Schrödinger E (1926a) Quantisierung als Eigenwertproblem (Erste Mitteilung). AP 79:361–376

Schrödinger E (1926b) Über das Verhältnis der Heisenberg-Born-Jordanschen Quantenmechanik zu der meinen. AP 79:734–756

Schrödinger E (1926c) Der stetige Übergang von der Mikro- zur Makromechanik. NW 14:664–666

Schrödinger E (1927) Der Energieimpulssatz der Materiewellen. AP 82:265–272

Schrödinger E (2003) Collected papers on wave mechanics. American Mathematical Society, Providence

Schweber SS (1990) The young John Clarke Slater and the development of quantum chemistry. HSPBS 20:339–406

Servos JW (1990) Physical chemistry from Ostwald to Pauling: the making of a science in America. Princeton University Press, Princeton

Serwer D (1977) Unmechanischer Zwang: Pauli, Heisenberg, and the rejection of the mechanical atom, 1923–1925. HSPS 8:189–256

Seth S (2010) Crafting the quantum: Arnold Sommerfeld and the practice of theory, 1890–1926. MIT Press, Cambridge

Small H (1986) Recapturing physics in the 1920s through citation analysis. Czechoslov J Phys 36:142–147

Sommerfeld A (1915) Zur Theorie der Balmerschen Serie. In: Sitzungsberichte der Bayerischen Akademie zu München, pp 425–458

Sommerfeld A (1942) Zwanzig Jahre spectroscopischer Theorie in München. Scientia 72:123–130

Sommerfeld A (1984) In: Eckert M, Pricha W, Schubert H, Torkar G (eds) Geheimrat Sommerfeld. Theoretischer Physiker. Eine Dokumentation aus seinem Nachlaß. Deutsches Museum, München

Sommerfeld A (2000–2004) In: Eckert M, Märker K (eds) Wissenschaftlicher Briefwechsel (2 vols). GNT-Verlag, Berlin (cited as SWB)

Somsen GJ (2008) A history of universalism: conceptions of the internationality of science from the Enlightenment to the Cold War. Minerva 46:361–376

Staley R (2005) On the co-creation of classical and modern physics. Isis 96:530–558

Walker M (2012) The 'national' in international and transnational science. Br J Hist Sci 45:359–376

Warburg E, Rubner M, Schlick M (eds) (1922) Helmholtz als Physiker, Physiologe und Philosoph. Müllersche Hofbuchhandlung, Karlsruhe

Widmalm S (1995) Science and neutrality: the Nobel Prizes of 1919 and scientific internationalism in Sweden. Minerva 33:339–360

Wise MN (2011) In: Carson et al (eds) Forman reformed, again, pp 415–431

Index

Printed in the United States
By Bookmasters